Space and Society

Series editor

Douglas A. Vakoch, SETI Institute, Mountain View, CA, USA
and California Institute of Integral Studies, San Francisco, CA, USA

More information about this series at http://www.springer.com/series/11929

Charles S. Cockell
Editor

Dissent, Revolution and Liberty Beyond Earth

Springer

Editor
Charles S. Cockell
UK Centre for Astrobiology
School of Physics and Astronomy
University of Edinburgh
Edinburgh
UK

ISSN 2199-3882 ISSN 2199-3890 (electronic)
Space and Society
ISBN 978-3-319-29347-9 ISBN 978-3-319-29349-3 (eBook)
DOI 10.1007/978-3-319-29349-3

Library of Congress Control Number: 2016932328

Cover design: Paul Duffield

Printed on acid-free paper

This Springer imprint is published by SpringerNature
The registered company is Springer International Publishing AG Switzerland

Preface

Faced with a political decision, a law or even an entire government that we find disagreeable, how are we to dissent? The question has perplexed social scientists for a long time and the resolution to this question has been extraordinarily diverse. From a minor and non-violent act of civil disobedience (such as refusing to pay taxes) to a full-scale geographically widespread bloody revolution, humanity has embarked on a dizzying array of methods to change 'the system'.

However, it is not clear if these methods (or at least the ones we might decide are desirable) can simply be transplanted into the space frontier; for example, a revolution that destroys infrastructure and causes depressurisation might kill everyone. Consequently, it is necessary to embark on a new discussion on the nature of dissent in space. How can we go about disagreeing with, and changing, the structures of governance put in place? How can liberty be preserved in the process?

On 11 and 12 June 2015 we continued the third and final discussion on extraterrestrial liberty begun by the UK Centre of Astrobiology and the British Interplanetary Society in 2013. It focused on the means by which dissent can be organised in outer space. It built on the previous two volumes resulting from our meetings on extraterrestrial liberty. The first volume examined the more general idea of what liberty is beyond Earth and what conditions might be necessary for liberty to survive in the extreme conditions of space (Cockell 2014). In the second volume, we took these concepts of liberty and gave them form by considering in more practical terms governance structures in space and how they might influence the type of liberty experienced by people in space (Cockell 2015).

This third and final volume in our trilogy on extraterrestrial liberty considers how we might rebel against the very social and governance structures we have created in space. How can we disagree with or dismantle organisational structures that no longer serve the purposes they were intended for?

We, the authors, would like to thank the British Interplanetary Society for supporting the discussion that has led to this collection. We would also like to thank Ramon Khanna and Alessia Valdarno at Springer and Doug Vakoch at the SETI Institute for bringing this book to fruition.

As has been said in the prefaces of the previous two volumes, the trilogy we have created has two functions. First, it is a set of essays written with the intention of contributing to a new branch of political philosophy concerned with extraterrestrial liberty. Second, these essays are a record of some of the thoughts of people in the twenty-first century who have never lived in space, but with the benefit of the expansive literature on liberty developed on Earth, the authors offer their ideas and thoughts on how liberty might develop among permanent denizens of the space frontier. At the end of this exercise, all of us who have taken part in these discussions and written these essays have only one message for those on the space frontier: good luck with your efforts to build spacefaring societies in which liberty can flourish!

Edinburgh
2015

Charles S. Cockell

References

Cockell, C. S. (Ed.) (2014) *The meaning of liberty beyond earth.* Berlin: Springer.
Cockell, C. S. (Ed.) (2015) *Human governance beyond earth: Implications for freedom.* Berlin: Springer.

Contents

Chapter 1
Introduction: Dissent, Revolution and Liberty Beyond Earth

Charles S. Cockell

Abstract Faced with an instantaneously lethal environment, it would be easy to believe that dissent would stand no chance in space. People in extraterrestrial settlements will become conservative since radical political, economic and technical change may threaten their existence. The authorities that run settlements, if nothing more than for their own credibility, will err on the side of caution and seek to quell dissent that either leaves them powerless or implicates them in a failure of leadership if the results of dissent lead to catastrophic structural failure. However, dissent must occur if extraterrestrial settlements are not to degenerate into places occupied by slaves doing the bidding of private or state enterprises. In this collection of essays, we examine how dissent and disobedience may manifest on the space frontier and suggest ways in which dissent may be allowed, even encouraged, to further political discussion and discourse in space. We examine how the conditions for dissent can ultimately influence the conditions for diverse forms of liberty in space.

Keywords Liberty · Dissent · Revolution · Government · Laws

'I like a little rebellion now and then. It is like a storm in the atmosphere.' So said Thomas Jefferson in 1787 in response to Shay's Rebellion. But can rebellion be tolerated in an environment where destruction of crucial infrastructure could deny people the air they need to breathe?

The extreme conditions of outer space that mandate collective efforts in survival and resource acquisition do not on the face of it lend themselves to free-thinking dissenters who want to do things 'their way'. Nor do they seem very conducive to the sort of civil disobedience that might be needed to change laws imposed by a despotic corporation or government agency charged with ensuring the survival of a group of people in outer space.

C.S. Cockell (✉)
UK Centre for Astrobiology, School of Physics and Astronomy, James Clerk Maxwell Building, the King's Buildings, University of Edinburgh, Edinburgh EH9 3JZ, UK
e-mail: c.s.cockell@ed.ac.uk

C.S. Cockell (ed.), *Dissent, Revolution and Liberty Beyond Earth*,
Space and Society, DOI 10.1007/978-3-319-29349-3_1

Having explored the general conditions for liberty in space (Cockell 2015a) and the influence of governance structures on that liberty (Cockell 2015b), we are left with the final questions: how are people to dissent from their structures of governance if they disagree with them? How can the freedom to dissent be realised in enclosed isolated outposts in space?

In this volume of essays, we take on this task, considering from a number of directions how dissent and revolution can be managed in space if they occur and how, on a more general level, healthy political and economic dissent and discussion is to be encouraged. As with our previous volumes, the authors come from diverse backgrounds, for example social and political scientists, planetary scientists and science fiction writers. It is this diverse set of viewpoints that provides us with a rich seam of ideas to examine how liberty will be expressed in space. Examining the future of liberty in space requires a quite disparate set of information. It will be inextricably linked to the environment, in the sense that the extreme conditions on any planetary body or in the open plateau of space will influence the political and economic conditions for liberty, for example by causing economic isolation and mandating political structures to ensure the safety of inhabitants. Thus, we need planetary scientists to bring knowledge of the physical conditions to be found in space.

To take this information and place it into the classic context of liberty—the centuries of existing discourse that started with the ancient Greeks and found particular strength in the literature of the Enlightenment—requires social scientists and political philosophers to bring their knowledge into the fray. It is pointless, and wrong, to consider the frontier of space some utterly new place for liberty when liberty will evolve in ways probably familiar to us in many respects and at least in ways linked to our past experiences of attempting to pursue liberty on Earth. Thus, this strong heritage of ideas must be harnessed.

Furthermore, although the discussion on liberty as a branch of political philosophy is new, liberty per se has not escaped the attentions of science fiction writers. To construct imaginative scenarios for human settlements beyond Earth, these writers have been forced to consider institutional arrangements and, by extension, the consequences of revolution and dissent in space. Science fiction is an extraordinary source of information for looking at various alternative futures and trajectories of liberty in space.

Thus, as with our previous volumes, this set of essays takes these various influences and here offers a set of ideas on how dissent could occur in space and what its major characteristics might be. We offer it as another contribution to the branch of political philosophy that deals with extraterrestrial liberty.

Dissent in space involves an adjudication of a range of competing interests. In a chapter that examines the challenge of how dissent might emerge in space, Tony Milligan investigates the range of competing interests in space that might complicate the extent and type of dissent that can be expressed. In particular, he suggests that the interests of future generations must also be considered. These differing interests have implications for the sort of political systems that might optimally support them while also allowing for disagreement. Constrained dissent, he suggests, might lead to a type of consensus politics whereby dissent acts as a temperature gauge to moderate the political process.

Charles Cockell explores the arena for civil disobedience in space. He suggests that anyone should wish to curtail violent disobedience and dissent because of the potential catastrophic consequences of the destruction of infrastructure, for example damage that might cause widespread depressurisation. The question then remains: how is anyone to rebel? Cockell discusses other courses of non-violent civil disobedience and suggests that many of the methods used on Earth could be applied in space. Nevertheless, the central point of his essay is that the extreme conditions of space make it easy for authorities to crush dissent in the interests of the safety of the people. Thus, active attempts must be made to allow for civil disobedience. One way to do this may be to establish a set of general rules or codes of behaviour, particularly those dissuading destructive violence, that would thereby explicitly allow for civil disobedience but establish boundary conditions based on civility in the extraterrestrial environment.

It would be easy to use the excuse that the extremity of the space environment requires dissent to be prevented at all costs, and certainly we might expect future authoritarian settlements to use this as an excuse to quell discord. A fundamental question therefore emerges: to what extent is dissent manageable and how much dissent should be allowable in confined space? James Schwartz explores the problem in the context of the Moon and looks at labour relations in lunar settlements. He concludes that strike action should be allowable, using analogies on Earth to bolster his case. His essay more generally underscores the importance of debating the extent of extraterrestrial liberty and which acts of disobedience should be protected and which not. His arguments apply equally to Mars and other locations beyond Earth. As labour relations underpin the entire economic and political functioning and success of any settlement, his essay homes in on one of the most important debates in extraterrestrial liberty: to what extent should the workers have the freedom to direct or withhold their own labour?

As in strike action, peer pressure has a large influence on the way in which people perceive their capacity to dissent or behave in ways not in accordance with a collective. Andrew Thomas examines some aspects of power and discipline in an extraterrestrial settlement and shows how the perception of being an outsider may have a powerful role to play in influencing whether people are willing to dissent. This problem, which touches on the 'tyranny of the majority', is made acute by the confining physical spaces of extraterrestrial settlements and the forms of discipline that emerge there. The chapter focuses the reader on the necessity to consider how power and discipline might develop in settlements and how it is to be managed and directed in constructive ways that do not unnecessarily undermine the freedom of the individual.

One clear way to understand the possible motives for dissent and how it is to be managed is to look at real examples. In a chapter that explores the precedent for mutiny in space, Mukesh Bhatt first looks at what constitutes dissent and mutiny and then explores examples of how this has occurred on space stations. He examines how existing legal structures, such as the Antarctic Treaty, might give us insights into how mutiny is to be managed and explores examples from fiction. This chapter touches on an aspect that is clear in other chapters as well: dealing with extraterrestrial dissent and attempting to maximise liberty in more permanent extraterrestrial settlements will not be an entirely new endeavour. There are a wide variety of experiences, including modern space exploration experiences, to effectively manage dissent and disagreement in space.

One of the most deleterious forms of dissent is terrorism—perhaps the most extreme and dangerous forms of dissent in confined, pressurised spaces. John Cain embarks on a discussion of space terrorism. He discusses both the means by which terrorists might go about trying to accomplish their goals and the ways in which they might be prevented. These countermeasures include punishment and legal means to discourage it in the first place. This chapter forcefully reminds us that with all the best planning and thought, we cannot prevent a rogue actor or actors from seeking to cause mass destruction. A settlement can be carefully engineered to maximise liberty, but this only works for the population willing to comply. Any consideration of extraterrestrial liberty must take into account those individuals, who, in seeking dissent through terrorist activities, will disregard any existing political, legal or even physical infrastructure. As terrorists could cause enormous damage in isolated space settlements, this unsavoury aspect of human disobedience must be considered as an important field of investigation in extraterrestrial liberty.

Stephen Baxter takes up the problem of how to allow dissent without catastrophic consequences. Specifically, he explores how the architecture of space settlements might be designed to allow dissent, even revolution, drawing on a rich legacy of science fiction in which these concerns have been raised. In particular, technological solutions, such as rovers that can be easily used to travel and effective methods to access resources in space, are ways in which people may derive the freedom to move around and leave settlements, reducing the chances that they will fall prey to despotic zealots in one location. His general premise—that there are technological methods to enhance dissent and thereby freedom—suggests that even prior to the large-scale colonisation of space, we can engage in engineering designs to maximise the opportunity that people have to establish free economic and political systems in which dissent is tolerated.

Developing a similar line of thinking, the essay by Lewis Pinault looks at how liberty and dissent might be engineered into societies beyond Earth, in particular looking at modern space endeavours and the movement of commercial enterprises into the space sector. In many ways, the efforts by national space agencies, such as NASA, to partner with commercial enterprise in the exploration of space already represents the changing patterns of liberty in space. When these new organisations disagree with existing agencies or try to change their policies, then they are already actively engaged in dissent in the space frontier—even if vicariously implemented from Earth in many cases at the present time.

Another way in which dissent can be expressed is through art. Art has the capacity to change our view of the world around us, but it can also be used as a channel through which to express revolutionary ideas. Drawing from her own experiences, Annalea Beattie explores the role of art in fashioning freedom beyond

Earth and its possible contribution to dissent. In the lethal extremes of space, art may take on an importance greater than on Earth as it provides a way for revolutionary ideas to find expression without physical disruption. It can become a means of quiet disobedience. We might even encourage all the various forms of art, from painting to theatre, as a way to neutralise, in a non-destructive way, dissent in extraterrestrial settlements.

Educating people in extraterrestrial settlements to be able to express their capacities and maximise their freedom, but at the same time to understand that they live in an environment that requires a sense of collective responsibility, will be crucial to the long-term success and sustainability of these settlements. Janet de Vigne explores an extraordinarily important part of the human lifespan: the teenage years. Full of angst, uncertainty and hormonal changes, how do we manage and educate the teenagers in a confined space where youthful errors allowed on Earth might threaten everyone in space? How do we achieve this without creating people who harbour a deep resentment against the settlement for what they perceive to be their restricted teenage years, yet the restrictions were necessary for the survival of the population? This chapter explores the multiple issues that underpin the education and channelling of the teenage years in space and how we might approach, in a systematic way, the education of teenagers.

When all else fails, one way to dissent is to flee. But this option is not always open to people and on the geographically confined surface of a sphere, i.e. Earth, movement may be possible but easy for governments to prevent. The vast expanses of space offer the possibility of flight as a means of dissent, but of course this capacity can only be realised if people have the means, such as spacecraft, to escape as well as another settlement to go to. Paul Rosenberg addresses the importance of flight in dissent and concludes that being able to flee is fundamental to our freedom. If his conclusion is correct, then we might even try to actively find ways to maximise the potential to flee in space to dilute the capacity for dictatorial control. Clearly, the vastness of space is a good start, but for people to be able to use this expanse to flee from autocrats, then we might consider constructing reliable trade routes and transport between settlements and, in the long term, maximising the number of settlements that exist so that people actually have the option of fleeing to somewhere else. In the early stages of human space settlement, fleeing may not be possible because of the limited extent of infrastructure beyond Earth, but as the effort progresses, the capacity for flight may improve.

Dissent can occur from individuals, but it may also occur at the scale of entire colonies. Kelly Smith investigates, using historical examples, how settlements may have imposed the conservative norms of individuals' founding mother country or world upon them, but eventually, in adapting to their environment, they generate new types of cultures that not only benefit the people but also the founding culture. Thus, dissent beyond Earth, particularly in establishing larger-scale novel patterns in culture, may be beneficial for everyone. His chapter shows that we should not merely consider dissent to be some sort of antagonism, but when practised at the scale of entire settlements, it may provide benefits for the whole of civilisation.

Continuing this theme of dissent at the scale above the individual, we might consider how the space economy could rebel against Earth by establishing a new direction for its development. Andrew Kennedy examines how the preservation of Earth can be made compatible with the exploration and settlement of space and explores models of how the space economy might be developed so as to become independent and successful. This includes the development of its own currency. The chapter is not so much a discussion of dissent and revolution within a settlement, but its underlying thesis is relevant to the way in which space settlements, and ultimately the entire space economy, might dissent from previous modes of economic and political development on Earth and establish its own trajectory.

1.1 Conclusion

In conclusion, this final volume in our series on extraterrestrial liberty completes a three-pronged approach to the political philosophy of freedom beyond Earth. In the first volume, we examined the conditions for liberty beyond Earth including the philosophical and practical foundations for understanding how liberty might develop beyond Earth. In the second volume, we transitioned into considering the political and economic implications of how liberty might manifest beyond Earth, in other words the implications for governance. In this final volume, we discuss what happens when those governance structures are not perceived to be serving the purposes and needs of an extraterrestrial settlement. The manner in which people and entire settlements can dissent and disagree, even to the point of revolution, is explored. None of these volumes on extraterrestrial liberty is exhaustive. Our objective was to contribute to a new branch of political philosophy concerned with extraterrestrial liberty. However, the essays presented here explore many aspects of this emerging discussion and should provide a solid foundation for further ideas on the future of liberty as humans continue their exploration and settlement of the space frontier.

References

Cockell, C. S. (Ed.). (2015a). *The meaning of liberty beyond earth*. Berlin: Springer.

Cockell, C. S. (Ed.). (2015b). *Human governance beyond earth: implications for freedom*. Berlin: Springer.

Chapter 2
Constrained Dissent and the Rights of Future Generations

Tony Milligan

Abstract The following paper will defend the viability, and constructive importance, of dissent within stable and more or less autonomous space settlements. That is to say, settlements of the sort which will come after the establishment of an initial human presence. (Which may have to be carried out under some form of command structure of a sort familiar from existing space programs.) The paper will, however, locate dissent and the entitlement to dissent within a broader cluster of commitments, placing it in tension with various other goods and duties which we are properly concerned to secure and fulfil. The politics of dissent in space are likely to be the politics of balancing out competing concerns. Specifically, it is assumed that some of us have a duty to try and help extend the presence of humans to nearby regions of space; that we also have duties towards future generations (and cannot inflict intolerable conditions upon them; and that we are and ought to be committed to various broadly-liberal freedoms including those concerning dissent). These concerns are individually plausible, difficult to abandon and also difficult to balance. It may be difficult to try to meet our duties without neglecting the legitimate interests of those who come after us and who cannot simply be sacrificed, in an illiberal manner, to our goals. Given that scope for dissent is basic to the freedom of any future generation, but also potentially dangerous, what seems to be important is that dissent of at least some kinds is (up to a point) both tolerated and constrained. The paper finishes with some comments upon the kind of political system which might be able to cope with the arising task of balancing out competing claims.

Keywords Dissent · Duty · Future generations · Authoritarianism · Liberal norms

What follows will suggest that there are some things which, surprisingly, we can and ought to do in space. Things such as engage in dissent: strikes, occupations, public demonstrations, non-cooperation, civil disobedience and (more generally)

T. Milligan (✉)
Department of Theology and Religious Studies, King's College London,
22 Kingsway, London WC2R 2LS, England, UK
e-mail: Anthony.milligan@kcl.ac.uk

C.S. Cockell (ed.), *Dissent, Revolution and Liberty Beyond Earth*,
Space and Society, DOI 10.1007/978-3-319-29349-3_2

acts which publicly challenge those in authority and their right to dictate policy or to command allegiance. This is surprising because it would involve dissent in a life-threatening situation. The paper is also, in light of the very real dangers of disagreement and dissent under extreme conditions, at least implicitly concerned with what we cannot and ought not to do, i.e. engage in reckless actions of the sort which feature in various familiar instances of science fiction. Robert Heinlein's *The Moon is a Harsh Mistress* (1966) is a case in point. It depicts lunar colonists in revolt against the Earth, hurling rocks at the latter. This is something which (trivially and obviously) we ought not to do and something which others also should not do. What I am proposing, instead, is a moral license for what we might call 'constrained dissent'.

As a point of clarification, there is nothing specifically novel or second-rate about a discussion of dissent as constrained rather than unconstrained. There are, after all, traditions of dissent which stretch to the violation of the law, but are in various respects *limited* by a concern to avoid harm. One obvious example is industrial action by members of the emergency services. (A problem explored elsewhere in this volume by James Schwartz.)[1] Another is civil disobedience, understood in a broadly Gandhian sense as protest which abides by non-violence and by civility towards the other. (A problem explored in the paper by Charles Cockell.)[2]

Constraint, in this Gandhian sense, does not imply passivity or half-heartedness but rather a determination to protest while avoiding certain sorts of dreadful, unintended, consequences or a spiraling out of control. (There are occasions on which a temporary setback or defeat is simply better than the likely outcome of pressing on regardless.)

In at least some respects, a non-destructive dissent in space would have to be constrained in a broadly similar sense. Everything could not be permissible, everything could not be possible. Revolution would be an extremely risky business, a rare and possibly catastrophic exception within a broader more routine pattern of dissent. It is dissent of the latter sort that I will focus upon. Such dissent would have to be tailored to the restrictive circumstances of life in a lethal environment. What follows will explore the bounds of such dissent and the way in which the constraining of dissent on the part of future generations might lead us to regard at least some settlements as ethically flawed projects, as projects which (all other things being equal) should not be undertaken. The attempt to say something about the permissibility and socially useful role of dissent will also point towards broader political requirements about the nature of the political system, the kind of democracy, which might work best elsewhere as a successor to the command structures of the first human settlements.

---

[1]See J.S.J. Schwartz, 'Lunar Labor Relations' in this volume.

[2]C. Cockell 'Disobedience in Outer Space', in this volume. For the Gandhian roots of this approach see Milligan (2014).

2.1 Picturing a Space Community

The initial human presence in a Moon base or on Mars will, no doubt, inherit the military-derived command structure of familiar forms of space exploration complete with mission commanders and the authorization of autonomous decision-making combined and ultimate answerability to agents back on Earth. The most appealing part of this picture is perhaps autonomy and it is likely to be part of this picture because of the difficulties of time-lags and the impracticality of distant decision making. We know this much from the Apollo program where the initial landing required a local over-ride of the targeted landing site in order to avoid mission failure. Even so, in combination with a degree of autonomy, a broadly military chain of command, stretching back to the Earth, is likely to be preserved. This is not, however, a political structure which would be indefinitely sustainable in the face of a more expanded human presence anywhere except perhaps on the Moon or in some sort of O'Neill habitat parked conveniently at a Lagrange point within easy reach of the Earth. The decisive issue is, again, proximity and practicality. Even on Mars, the pressures towards a preponderance of autonomy and of local control would be difficult, over the course of time, to resist. At some point, any settled community is likely to regard the Earth as too distant and lacking in an immediate grasp of local realities. The more distant the colony is from the Earth, the stronger the pressure towards local autonomy is likely to be. Yet autonomy is consistent with the shortening of lines of communication between elites, bureaucracies and security personnel. It is consistent with the local dominance of authoritarian and military-style command structures rather than anything resembling a democracy.

Nonetheless, from a settlement survival point of view and assuming a capacity for any settlement to be self-supporting (if necessary), the promotion of autonomous-control could be possible and could be the right thing to do. There is an obvious sense in which the hands-on practicalities of community building are likely to be understood far better by those in situ rather than by agents who are months away in terms of journey time and perhaps more distant still in terms of mindset. (Again, Mars is a better model here than the Moon, which is still nearer to us in journey time than the Americas were to the main seats of colonial power in early modern Europe.) This 'locals know better' claim does not, of course, apply to all knowledge or to all survival requirements. Sometimes those on Earth will have a better breadth of vision and will know what has to be done far better than a settler community. They may have a better appreciation of the bigger picture of what makes a community sustainable over the course of time. Arising from this, there are ethical issues in play concerning the disclosure of risk and levels of difficulty. There is information that authorities on Earth might sometimes be tempted to withhold in order to sustain trust, co-operation and morale. Yet there will still be at least some epistemic advantages which settlers will (almost unavoidably) have over agents who are situated within a terrestrially-based command system. Most obviously, settlers will have a better grasp of the terrain and the phenomenology of 'being a settler', a better grasp of 'what it is like' to experience such a situation. While the motivating rationale for settlement may include the interests of agents who remain emphatically Earth-bound, too much terrestrial resistance to settlement autonomy may be unwise for all concerned, although timing may matter. The outlook of those born into a settler community is likely to be very different from that of the pioneers, adventurers, misfits and escapists who are deemed to have the right stuff to join an initial settler body. Too rapid a transfer of control may be as bad as too slow a transfer. One upshot of this way of regarding matters is that it may be better to speak, in the above manner, of 'settlement' as the ultimate goal of an offworld expansion rather than 'colonization', given the answerability to a power elsewhere which the latter implies. 'Colonization' also carries unwelcome overtones of exploitation and dominance which may, similarly, be something of a bad start. 'Settlement' and talk of 'settler communities' may not be entirely free of negative associations however these associations are at least somewhat weaker.

In what follows, the focus will not only be upon routine dissent, but upon such dissent within a particular type of context, the type of context supplied by more or less autonomous and stable non-terrestrial communities of a scale which is greater than that of any initial settlement party and, a little more precisely, beyond the scale on which an initial quasi-military command structure would be either appropriate or adequate. We might, of course, argue about exactly how large the population would have to be before this was the case and somewhere in such a discussion appeal might be made to the thousands-strong communities of humans already living in the polar extremes on Earth. The population of Antarctica is about 4000 throughout the year with about 1000 people wintering over. However the fluctuating numbers of such communities indicates that they are not communities of the right, more or less stable, sort. Stable communities will, instead, have a concern for social reproduction, for the reproduction rather than the replenishment of population from elsewhere; for the local reproduction of the means of existence (assuming the self-sufficiency specified above); and for the reproduction of various social goods, such as solidarity, respect and mutual recognition as well as any material goods which might be required to avoid the alienation associated with a strong sense of deprivation and inequality of circumstance.

I will take it as important (for survival purposes and not simply morale) that members of such settlements do not feel themselves to be, in relation to the Earth, permanently at the edge of the fun fair but never quite able to join in the festivities. In short, for the sake of sustaining a sense of social hope, some approximation to a good life must be possible. Social hope of this sort, although it has been defined in various different ways (by philosophers such as Richard Rorty, social theorists such as Manuel Castells, and political figures such as Vaclav Havel), seems to be an important factor in convincing agents that some sacrifices are necessary for the common good and that the common good itself is the good of a community which is actually worth safeguarding (Rorty 1998, 1999; Castells 2015; Havel 1991). And while hope and sacrifice may also be consistent with penal settlements, in which compulsory labour is followed by freedom and citizenship, what I will assume here is that settlement membership will not be either a form of punishment or an idealistic sacrifice for the sake of mankind (and not a foolish mistake by those who do not realize what they are getting). What is envisaged is neither a penal community nor a saintly community, and certainly not a ship of fools, but rather a modestly sized ordinary, stable community of people living out their lives under extraordinary circumstances. People who, moreover and by virtue of their ordinariness, will be inclined to complain about their lot when they regard it as in some way intolerable or at least unjust.

Finally, in line with previous work (Schwartz 2015; Milligan 2015b), I will assume that such deliberation need not be mere conjecture and may reasonably be constrained by appeal to a Rawslian 'veil of ignorance' test for extraterrestrial political systems (Rawls 1971). In brief, to consider whether a candidate proposal or approach to political organization is at all viable, we should consider whether or not we would agree to it under circumstances where we happened to be unaware of how it would impact upon us. While this might not be the best approach to terrestrial political arrangements (because we know already happened to know what it is like to be terrestrial agents, how we are actually situated in the world and what would and would not constitute terrestrial justice) it is an approach which has a good deal to offer in relation to deliberation about the novel (as yet unlived) circumstances of life in space. Until we have a better constraining strategy for deliberation, a variant of the Rawlsian 'veil of ignorance' may offer the best test of the *fairness* of any extraterrestrial political theory.

2.2 The Problem of Competing Goods

Our attitude towards any such stable but ordinary community faces something of a difficulty, beyond lack of experience concerning actual space settlement. The difficulty arises because a number of our more plausible ethico-political commitments concerning power, rights, dissent and the state, turn out to be in tension with one another. Such tension is not, however, a sign of any obvious adherence to one-commitment-too-many or mistaken allegiance to a belief which need only be removed in order to secure a tension-free consistency. Rather, I want to suggest that tensions of the relevant sort tell us something about what it is to be human and consequently to value different goods at one and the same time. Elsewhere, I have suggested that space ethics is, in certain irreducible respects, dilemmatic for precisely this reason (we are naturally and perhaps inescapably pluralists when it comes to goods). This is a claim which I would like to extend now to ethico-political matters. More specifically, the following three claims identify different sorts of duties which go hand in hand with the goods of human survival, the opportunity of a good life for future generations and the goods associated with liberal freedoms. While they are individually plausible, when brought together they generate difficulties:

(1) Some (not all) currently-existing agents have a duty to try to extend human life through space settlement.
(2) Assuming that we are agents of the sort specified in (1), we are also agents who have duties towards future generations (wherever situated).

And

(3) We are and ought to be committed to the survival of at least some basic liberal political norms, such as those concerning the permissibility of dissent.

These claims do not generalize in a way which places an impossibly demanding burden upon any humans. Those who are extremely poor and, to all intents and purposes powerless with regard to consciously exerted influence upon humanity's future, cannot reasonably be said to have duties of the sort specified in the first two claims. Only some of us can have duties of the relevant sort and the assumption here is that some of us do indeed have such duties. However, this fact alone does not remove the tension between these claims. Nor it is obvious that such a tension should always be resolved in favour of fulfilling the duty which is specified in (1) at the expense of the liberal norms alluded to in (3). Indeed, one of the running themes of the Extraterrestrial Liberty discussions out of which this text has emerged is the suitability of deliberating about short and medium term space objectives from a broadly liberal and pragmatic standpoint rather than from a standpoint which is dominated entirely by a sense of the need for survival and by what Stephen Baxter has referred to as 'the cold equations' of some kind of neo-Darwinian survivalist ethic, an ethic which would simply be unlivable over any protracted period of time for most human agents (Baxter 2015).

Additionally, some space settlements (even if they could be stabilized) would not obviously be worth establishing even if doing so helped to further the duty specified in (1). As an extreme example, a settlement which would require or collapse into a dictatorially abusive junta-dominated system does not seem to be an ethically worthwhile goal. Solving the justification problem for space settlement (explaining why it might *sometimes* be worth the required resources) is a difficult task. Doing so for some manner of 'Iron Skies' option is likely to prove impossible. Here, I allude to a popular, and rather lightweight, science-fiction film about a Fourth Reich on the Moon and also to Jack London's classic dystopian novel of resistance to tyranny, *The Iron Heel* (1908). But even without the introduction of space-juntas, jackboots or tanks upon the lawn, more common or garden curtailments of freedoms may be anticipated as a real possibility. Abortion rights as well as freedoms to carry to term without severe penalties may be compromised in the interests of community sustainability depending upon the carrying capacity of a local settlement and the difficulty of becoming pregnant and producing offspring under non-terrestrial conditions (Milligan 2015b). More generally, if a space settlement is too liberal in its freedoms and entitlements to dissent then it is unlikely to survive. But if it is too illiberal then establishing it might not be a worthwhile or defensible goal (and, for such a strongly-illiberal settlement, survival problems are also likely to emerge over the course of time, for reasons specified below). Beyond a certain point, we may well decide that the exercise of settling some location *L* may be too likely to compromise our understanding of basic liberal norms.

Admittedly, some of these comments take us into a very grey area but, in a sense, that is the natural element of liberal political discourse and it gives no immediate reason for us to retrace our steps in order to look for a better way to smooth out all tensions or to reconcile all competing commitments. What we can say with some confidence is that an entitlement to dissent is basic to liberal political norms, basic to stability and to social solidarity and basic also to respect for persons. So too is an entitlement to *tolerant response* in the case of protest which, even if sometimes illegal, abides by reasonable standards of civility such as the avoidance of reckless endangerment. Legal penalties may be expected in the latter case, but they should differ in terms of stigma and severity from those applied in response to more commonplace forms of lawbreaking and in response to life-threatening political violence (such as terrorism or the use of revolutionary political force). A difficulty still arises here because it is by no means obvious that this approach *can* be put in place for *all* extraterrestrial communities in *all* places and at *all* points in time. Dissent which might have no great detrimental terrestrial impact on Earth could, conceivably, yield a Hobbesian meltdown in at least some contexts elsewhere, given prevailing conditions of extreme vulnerability. What does not constitute reckless or life-threatening action *here* might constitute both *elsewhere.*

And two further points make matters even worse. Firstly, a point stressed on several occasions by Charles Cockell: the conditions of vulnerability and dependence in space will generate a tendency towards authoritarian political control within any settlement. Whoever controls the air, food and water will be in a position to control the people (Cockell 2013). Secondly, a point which owes something to Mikhail Baktin's Classic study of popular early modern carnival and dissent: authoritarianism creates its own counter-culture (Bakhtin 2009). This claim seems to be borne out by any reasonably detailed examination of human history at any historical period. Dissent, like the poor, seems always to be with us. It is not something we can reasonably wish away even if we wanted to do so. It is not something which could be eradicated from communities of the sort specified above, even through the use of determined authoritarian measures and even though the dissenting agents will be acting in the life-threatening environment of space. Accordingly, even if it was morally permissible to support an Iron Skies option of radically authoritarian political control, such an option would not actually eradicate the pressures towards dissent and it might, because of its extreme inflexibility, prove to be ultimately counterproductive. Rather than imposing obedience by main force, it might generate problems which could well cut across the fulfilment of the duty specified in (1) to extend human life.

2.3 A Closer Look at the Commitments

The first two commitments are broadly communitarian, the third is more ambiguous and may be read in either an individualist or a communitarian manner, as a concern for individual rights and entitlements or as a concern for collective entitlements such as the right to strike, to form and join trades unions, and to protest or speak out against the government. Upon closer examination, all three commitments have a degree of ambiguity about them. The idea of extending human life in (1) can be read in at least two quite different ways, as a claim about the preservation of our species or as a claim about the preservation of our 'moral community' (the bounds of which are not necessarily set by common biological origins or shared patterns of DNA). The idea of duties to future generations, cited in (2), can also be read as an obligation to particular future agents or to future groups of humans no matter who they turn out to be. In line with a previously supported position (Milligan 2015a) I will suggest that we resolve the former ambiguity by endorsing the idea that our primary obligation to humanity is to a moral community (rather than a species) and that we resolve the latter ambiguity by accepting that our duty to future humans is, in some sense, to a collectivity rather than to particular individuals.

One important reason for the former move is that a species-focused reading of our duty to humanity generates too many anomalies. For example, if it were discovered that a section of the population were 'Spock-like' and did not actually share human DNA would this in any way entitle us to exclude them, or their descendants, from our survival plans? Doing so would look suspiciously like a form of prejudice as well as being practically unwise. By virtue of their biological difference they might turn out to be better equipped to survive and to preserve the legacy of our shared community. Moreover, it is the species reading of a duty to extend human life which is connected to the kind of narrow and impractical consequentialism which holds that *any* sacrifice is legitimate if it is required to secure survival. The moral community reading of our duty to humanity is, contrastingly, tied to the possibility that through certain kinds of moral failures a community might (figuratively) 'forfeit its right to survive' or might undermine the value of community survival through a betrayal of the basic values on which the sense of community is based. In less abstract terms, it is better to be part of a short lived but good community rather than a long lived but morally bankrupt or even repugnant community. A Fourth Reich in space, no matter how long it lasted, could not count as success. With this understanding of our duty to humanity in place, we can begin to appreciate that while there is a tension between (1) and (3) there is also a commonality of concern. Together they combine into a commitment to survival under conditions which do not dehumanize but instead offer some manner of freedom, even if it is of the paradoxical sort referred to by Cockell (2015) as 'freedom in a box'.

The second commitment, concerning duties towards future generations, is like the first in the sense that it is a more obviously communitarian than individualist concern. Duties towards future generations really do seem to be duties towards generations and not towards specific individuals for reasons specified in what is known as the 'non-identity problem' (Parfit 1986). If action φ involves environmental degradation which might, in 100 years time, lead to harm to members of a future generation, *Bob*, *Mary*, *Elizabeth* et al. I might try to avoid it. But in doing so I am likely to alter who comes into existence (which may no longer be *Bob*, *Mary*, *Elizabeth* et al.) because who ends up coming into existence is sensitive to even minor alterations in our current behaviour. (This is also the reason why time travelers in popular television programs and films are warned not to change anything because they might then end up not being born.) I can do 'the right thing' for the sake of (a) future generations whoever their constituents are, and (b) those particular future agents who will come into existence no matter what I do (and the number of these will tend towards zero as we move beyond the next generation). But I cannot 'do the right thing' for anyone whose existence I cannot actually and reliably foresee. This, again, stresses the broadly communitarian nature of the duties in question. And this is no bad thing. The right kind of ethic for space may (I think 'should') acknowledge the value of individuals as unique and irreplaceable, but it is unlikely to be a strong form of individualism. Although not focused upon particular discrete known-in-advanced individuals, such future-oriented duties do seem to be real. As a symptomatic but extreme illustration: we do not get to plant time bombs in the public square just so long as no currently living and determinate or identifiable agents will be harmed. There is no requirement that the likely victims of our actions must enjoy the terra firma of existence in order for us to have duties to avoid the actions in question. This kind of future-oriented deliberation is familiar from discussions of nuclear waste management and environmental harm more generally (Routley and Routley 1978). In short, we do not get to ignore any likely and dire consequences of our actions even if currently existing agents will not live to see or be harmed by them.

The third commitment is more focused upon individual rights or at least upon a combination of individual and collective entitlements. And so the way of resolving its ambiguity is not as clear cut as opting for one thing or another as an important good. It certainly does not reduce simply to a liberal valuing of choice. My suggestion that a problem arises from of our duties towards future generations in space settlements is not the familiar one that it may be wrong to bring individual agents, involuntarily, into existence under the compromised circumstances of having to live somewhere other than in the midst of the lush greenery of the Earth. The latter claim may have something to it, but is problematic if pressed too far. After all, every human agent is brought into being under compromised circumstances which are also not of their own choosing. What matters is not so much choice as such but rather the nature of the (admittedly unchosen) circumstances. Those who think of space exploration in terms of a basic biological survival imperative may consider such issues moot because humans will breed no matter what ethicists say. But here we would do well to remember that not all humans will do so, that many couples in the West now choose to have only one child or none at all, and that slaves often tried to entirely avoid reproduction precisely because of the conditions under which any offspring would have to live. Yet here we may wonder about just how bad circumstances would have to be before we were justified in thinking about matters in the same way in relation to bringing agents into being in space. On a crude consequentialist account it might be tempting to say that we ought only to bring humans into being where there is likely to be a preponderance of happiness over suffering within their lives, or where the net balance is likely to be at least neutral. But what makes it difficult to buy into this view is that it would make most human reproduction, now and throughout history, a mistake. We need not lapse into some manner of philosophical pessimism in order to accept the inconvenient truth that life has generally been quite difficult for most humans at most times in the past and this remains the case in many parts of the world even today.

Rather than appealing to some dubious consequentialist claim (in which things, overall, shall be well) it seems more plausible to hold that whoever we bring into being should have at least the *opportunity* for some sort of good life (on a complex understanding of the latter) even if suffering figures as a component part of such a life just as it figures as a component part of our lives and most human lives. It should be a life, in short, which the agent themselves could readily accept as meaningful in spite of suffering and meaningful and worthwhile in its own right rather than being simply a part of someone else's grand plan. To say this much is precisely to regard such a life in terms which are familiar from a liberal discourse of respect, integrity and individual worth. It is also to appeal to a sense of inter-generational justice. And this is where a Rawlsian test seems to fit quite well. Under imaginary conditions where we do not know which generation we were going to belong to (the present generation or a future generation in a space settlement) we might well hold that enough had been done to offer a good life no matter where we happen to end up situated, and no matter when. This being so, the opportunity of a good life is the kind of requirement which might lead us to regard arrangements as just or as an approximation to justice. They might well pass a Rawlsian test for political deliberation about the unfamiliar conditions of life in space.

To deny basic freedoms and any entitlement to dissent, simply in order to fulfil our terrestrially-derived plans, when we are sure of actually living in better times than those who might come later, looks like a way of regarding future generations only as a means to our ends rather than as groups of beings who will have ends and aspirations of their own, beings who will (in a familiar philosophical terminology derived from Kant) be *ends in their own right* and beings towards whom justice is due. In the light of this, the case for upholding claim (3) in relation to space starts to look strong. Indeed, the case for deliberating about an orderly transition from first landing to at least some political freedoms and entitlements to dissent within a settled autonomous community, looks more or less clear cut. If we cannot look ahead to the realistic possibility of political freedoms then the project of settlement will involve, from the outset, a notable moral failure of duty towards future generations. We may, of course, become trapped at some point in a tragic dilemma where some form of wrongdoing is tolerable because it is unavoidable, because we face only a choice of moral failures. But we may also have a reasonable hope for something better and this goes to the heart of what it is to have 'social hope' of the sort which political justice ultimate requires.

2.4 Dissent and Political Structure

Even so, embedded rights and entitlements to dissent are unlikely to reproduce anything which will look closely similar to terrestrial liberty in a modern liberal democracy. Liberty is always likely to be compromised and constrained or boxed-in. And so, as conceded at the outset, violence of the sort which helped to bring various of the Western democracies into being (the American and French Revolutions, the revolutions of 1848, two major international wars and multiple militarized civil conflicts) are likely to be mutually destructive for the contending parties. In space, there can be no question of destroying and then rebuilding vital infrastructure at some later point in time. Rather, life support systems must be maintained and emergency measures of an authoritarian sort may sometimes have to be tolerated as the only way in which this can be done speedily and efficiently. And this may have to be accepted even though it is a circumstance which would probably be exploited in unfair ways by elites under the cover of acting in the general interest. And their position too is worthy of note.

While the focus of this volume has predominantly been upon the position of ordinary citizens or subjects of extraterrestrial political systems, it is worth pausing for a moment to consider the standpoint of elites on the assumption that the latter may well exist in the kinds of stable societies which are envisaged. We need not, of course, exclude the possibility of some more radically egalitarian option but we may have to exclude any reasonable possibility of arriving at it as the immediate successor arrangement following on from the command systems of initial settlement. The ideologically tempting option for elites faced with dissent is likely to be broadly Hobbesian. That is to say, the message that safety is simply too fragile to justify *any* defiance of the powers that be will appeal. When Thomas Hobbes penned *Leviathan* (1651), the classic defense of authority as outside of any social contract and beyond answerability to the population at large, he may have been wrong about the ever-present precariousness of safety, but his picture of a life without absolute, top-down, sovereignty tending to be ‘nasty, brutish’ and, above all, ‘short’ will, no doubt, be a tempting picture for those who would benefit most directly from authoritarian rule.

Nonetheless, the Hobbesian option is short-termist in a context where attention to the bigger picture of securing longer-term stability is required. I say this because many if not most settlements in space are likely to fail at some point in the face of external obstacles and internal weaknesses although the nature of failure may vary from continuing dependence upon the Earth through to actual starvation and system collapse. Of the difficulties making success so hard to achieve, the internal problems may at least be mitigated, but not entirely removed, by tapping into the socially advantageous side of dissent and into the collective wisdom which it often represents. There are, after all, matters which no individual should face alone for the simple reason that they are too large to fully appreciate. Our individual ethical sensibilities have been formed through coming to terms with much more restricted, localized and personal predicaments. Familiar ethical experiences have not equipped us with the kind of practical wisdom which would be required to individually come to grips with survival issues where the future of humanity as a whole is at stake. Shared deliberation, but also dissent and certain kinds of democratic constraining, may be ways of getting beyond this limitation and tapping into something more distributed and greater.

For this reason, although I have favoured Rawlsian deliberation and liberal political norms, there is a strongly communitarian dimension to the picture painted above and this feeds through into a valuing of dissent rather than a Hobbesian suspicion about the latter as a socially-destructive threat. Dissent would have to be a feature of any healthy political system for several reasons. The most obvious of these is the psychological constraint that agents cannot actually live in a constantly future-oriented mindset with every action shaped by a primary concern for humanity's future. The needs and desires of currently-existing agents must always be addressed as one of the goods towards which a functioning political system aims. And while it may be only one of the goods pursued by a stable and functioning state (because elite interests invariably do shape policy) it is still necessarily a part of the mix. The most likely outcome of, instead, promoting only elite interests or a continuous sacrifice for the future, out of a sense of moral idealism, would be a deep resentment against the state, against future generations and against those whose grandiose plans have generated this unfair predicament in the first place. (That would be 'us'.)

Alexander Herzen's classic 19th century warning about this kind of mentality is still striking. 'Do you truly wish to condemn all human beings alive today to the sad role of caryatids supporting a floor for others some day to dance on…or of wretched galley slaves, up to their knees in mud, dragging a barge filled with some mysterious treasure and with the words "progress in the future" inscribed on its bows?' (Herzen 1979, p. 37). Some years ago, as part of the anniversary of another rather well known, dystopian, text about the future, Glover (1984) reflected upon Herzen's warning by suggesting that such a heavy emphasis upon future people combined with persistent sacrifice for their sake would be not only unfair but would risk a constant deferral of benefits. These points still hold. Future-orientedness would not work in the present and there is no reason to believe that the future generations of agents who might live in a stable settlement would or should be willing to sacrifice so much and still sustain the levels of solidarity and commitment required to survive. A technologically advanced world without liberties, including certain kinds of dissent, is also likely to be a world without hope.

More broadly (and these are considerations which borrow again from Rawls) dissent seems able to perform (i) basic expressive functions; (ii) communicative functions; and (iii) a corrective function in the face of democratic deficits and political failures. In addition, borrowing rather more from the political experience of the early 21st century, dissent can be a corrective to the problem of intransigent majorities who are unwilling to make concessions in order to secure social solidarity. Of course, as Rawls points out, there is a danger that dissent, even in its most useful and constructive forms such as civil disobedience, can overwhelm a political system if it is badly timed: when, for example, broadly justifiable political institutions are already reeling under multiple sources of protest and discontent (Rawls 1971). Similarly, dissent can widen social divisions when those in authority dig their heels in and refuse to budge.

What this may steers us towards is not a pathological or Hobbesian hostility towards political protest and unrest but rather a conception of constrained-dissent tolerant political systems as the best candidate systems to move towards after an initial command-control settlement. And here, what counts a constrained-dissent rather than recklessness will vary from case to case and from settlement to settlement. (Just as there are variations in what these might involve when we contrast the Earth and anywhere else. What is reasonable here may not be reasonable elsewhere.) As for the more distant future, that really is a matter of conjecture, although we may doubt that anything akin to the utopia envisaged by some early space pioneers, such as Konstantin Tsoilkovsky will ever arrive. But with regard to a level of tolerance for dissent, we may still do rather more than suggest (or rely upon) an enlightened survival-oriented attitude on the part of ruling elites. Indeed, a reliance upon elites looks suspiciously like a losing political strategy under any circumstances and particularly so given their epistemically-compromised situation, their tendency to become enclosed, trapped in their own privileged word to the detriment of society as a whole. Rather, tolerance for dissent or more specifically for *certain kinds of* dissent is something which is best structured into the political process in ways which do not leave goodwill (a scarce resource at the best of times) to do most of the heavy lifting.

As a closing point, there is at least one kind of political process which attempts to do just this, one which has gained increasing attention in recent years both among protest movements (such as the Occupy movement of 2011) and among those aiming to instate peace processes (such as the process in Northern Ireland), i.e. consensus deliberation (Graeber 2014; Castells 2015). All other things being equal, if we are ever faced with the task of trying to build settled communities, there is a case for saying that we ought to be aiming and planning for a transition to something like consensus deliberation systems. In other words, democracy but not quite as we know it (at least with regard to the treatment of minorities). Consensus in the relevant specialized sense is not about getting an impossible 100 % agreement as to the *best* course of action, but rather involves arriving at an option which everyone, or almost everyone, and not just electoral victors, can live with. This approach contrasts strongly with the dominant versions of parliamentary democracy in which policy tends to be shaped by majorities (which are often majorities only in the sense of parliamentary outcomes rather than actual popular support) who give way to the Cromwellian urge of imposing their will. A constrained dissent, which acknowledges that the risks of widespread political violence are too great, and a consensus politics through which the authority of majorities (or of the larger of several competing minorities) is acknowledged but constrained by constitutional rights and minority vetoes, may even be potential allies. This is all very pragmatic territory with the rough edges of political monomania removed.

Constrained dissent, where the constraint is built into the very culture of dissent, and is not itself simply a matter of following the guide-rails established and tolerated by political elites, may not only be an (admittedly fallible) way of avoiding a violent political meltdown, it may also be a way of acknowledging the standing of majorities while providing an informal mechanism (in addition to formal vetoes and constitutional rights) to hold their power in check and to show the extent of feeling among non-dominant groups. Dissent, in this sense, provides a 'check' for emerging social divisions which might be addressed while there is still time, before they reach threatening proportions. If a popular culture of constrained dissent and consensus structures could be married together then there is a possibility that some version of democracy off-world might just be inclusive enough and stable enough *for long enough* to allow a chance of a good life for its citizens and a decent chance of community survival up to the point of settlement division. This could give at least some autonomous settlements a fighting chance of defying the odds by evolving the new and unanticipated political tools which will, no doubt, be needed if any such settlements are ever to survive under the unprecedented pressures of life in space.

References

Bakhtin, M. (2009). *Rabelaise and his world.* Indiana: Indiana University Press.

Baxter, S. (2015). The cold equations. C. Cockell (Ed.) *The meaning of liberty beyond earth* (pp. 13–31). Heidelberg, London, New York: Springer.

Castells, M. (2015). *Networks of outrage and hope: social movements in the internet age* (2nd ed.). Cambridge: Polity.

Cockell, C. (2013). *Extraterrestrial liberty: An enquiry into the nature and causes of tyrranical government beyond the earth.* Edinburgh: Shoving Leopard.

Cockell, C. (2015). Freedom in a box: paradoxes in the structure of extraterrestrial liberty. C. Cockell (Ed.) *The meaning of liberty beyond earth* (pp. 47–68). Heidelberg, London, New York: Springer.

Glover, J. (1984). *What sort of people should there be?* Harmondsworth: Penguin.

Graeber, D. (2014). *The democracy project.* London: Penguin.

Havel, V. (1991). The Politics of Hope. In *Disturbing the peace.* New York: Vintage.

Herzen, A. (1979). *From the other shore.* Oxford: Oxford University Press.

Milligan, T. (2013). *Civil disobedience: protest, justification and the law.* New York: Bloomsbury.

Milligan, T. (2014). Civility and politicized love in Gandhi. *Relig South Asia, 8*(3), 285–300.

Milligan, T. (2015a). *Nobody owns the moon: The ethics of space exploitation.* North Carolina: McFarland.

Milligan, T. (2015b). Rawlsian deliberation about space settlement, C. Cockell (Ed.) *The meaning of liberty beyond earth* (pp. 9–22). Heidelberg, London, New York: Springer.

Parfit, D. (1986). *Reasons and persons.* Oxford: Oxford University Press.

Rawls, J. (1971). *A theory of justice.* Cambridge: Harvard University Press.

Rorty, R. (1998). The end of leninism, Havel, and social hope. In *Richard Rorty: truth and progress: philosophical papers volume 3* (pp. 228–244). Cambridge: Cambridge University Press.

Rorty, R. (1999). *Philosophy and social hope.* London: Penguin.

Routley, R., & Routley, V. (1978). Nuclear energy and obligations to the future. *Inquiry, 21*, 133–179.

Schwartz, J. S. J. (2015). Fairness as a moral grounding for space policy. C. Cockell (Ed.) *The meaning of liberty beyond earth* (pp. 69–89). Heidelberg, London, New York: Springer.

Chapter 3
Disobedience in Outer Space

Charles S. Cockell

Abstract How are individuals and groups of people in extraterrestrial settlements to rebel against edicts, laws and even governments with which they disagree? The instantaneously lethal conditions in outer space make violent disobedience undesirable because of the potentially catastrophic consequences of the destruction of infrastructure. There are approaches, such as a free press and open political deliberations, that may discourage people from engaging in violent disobedience by providing them with effective channels for dissent. Furthermore, habitats can be engineered to mitigate catastrophic effects of violent disobedience should it occur. Additionally, there are various mechanisms of non-violent civil disobedience that could be used by a population to register dissent. By overtly recognising the need for mechanisms of civil disobedience within rules, dissent may have a chance of being included in productive political discussion.

Keywords Disobedience · Dissent · Liberty · Revolution

3.1 Introduction

The ability to dissent against unwanted governments, laws and state edicts is essential in any society where people are not to be cajoled into submissive conformity by dictatorial rulers. On Earth, mechanisms of disobedience have ranged from rather benign student 'sit-ins' in universities to more violent disobedience

---

C.S. Cockell (✉)
UK Centre for Astrobiology, School of Physics and Astronomy, James Clerk Maxwell Building, the King's Buildings, University of Edinburgh, Edinburgh EH9 3JZ, UK
e-mail: c.s.cockell@ed.ac.uk

C.S. Cockell (ed.), *Dissent, Revolution and Liberty Beyond Earth*,
Space and Society, DOI 10.1007/978-3-319-29349-3_3

such as the practice of tree-spiking by ecological dissenters.[1] Can these approaches and experiences simply be transplanted into outer space to allow for the same types of dissent beyond Earth?

It seems certain that they cannot. The environmental context is very different. Lethal external conditions make violent disobedience dangerous. The general culture of conformity generated by extreme environments will likely dull both the courage people have to be disobedient and the alacrity with which schemes and plans for disobedience are accepted by the population.

The interdependency between people in space instantly creates a major limitation to social disobedience. One cannot simply remove oneself from a society with which one has a disagreement; as Thoreau famously said (Thoreau 2010, p. 26), 'Know all men by these presents, that I, Henry Thoreau, do not wish to be regarded as a member of any society which I have not joined.' Now even on Earth, one might argue convincingly that few people can truly remove themselves from society. Even though we can breathe the atmosphere and, in some places, get access to water and food from the environment, most of us, even in this near state of nature, still depend on other people. However, on Earth, we are more able to approach a state of removing ourselves from a society than in outer space, where to even open the airlock as the first step of physically dissociating oneself from a society is to be instantly killed.

Thus, in outer space, disobedience can rarely, if ever, involve a declaration that one no longer belongs to the society of which one now disapproves. Disobedience must be accomplished from within the existing social structures of society and within the physical space and confines of pressurised habitats. Disobedience is, by the nature of the environment, trapped within a container, making the rulers potentially more powerful and the ruled potentially more easily cowed. The question is how disobedience is to be achieved at all within these conditions.

It is therefore essential to attempt to understand some of the problems for and limitations of disobedience in space and contemplate some of the ways in which it can be made possible without severely compromising the existence of a settlement.

In this essay, I first consider the environmental context of disobedience, which I contend is one of the major factors that constrains the scope and plausibility of disobedience,[2] particularly violent disobedience, as a means of dissent. I then consider some approaches that might be taken to minimise violent disobedience and

[1]The history of civil disobedience is a rich one and long before this phrase was coined it was considered under other phrases such as the 'right to resistance'. The concept that bad government should be overthrown was elaborated by Locke (1988), but others before and after him also explored the ideas of active resistance to bad laws and governments. This history is nicely summarised by Laudani (2013, pp. 9–31).

[2]Civil disobedience is, it should be recognised, possibly the first step in trying to reverse policies or dictates. As Zinn (2013, p. 16) points out, 'an act of civil disobedience, like any move towards reform, is more like the first push up a hill'. It can in some ways be regarded as a pressure valve for allowing dissent before it gets out of hand. Zinn later states the following (Zinn 2013, p. 108): 'That is exactly the point of civil disobedience, of a politics of protest—that it is an attempt to bring about revolutionary social changes without the enormous human toll of suicidal violence or

what engineering measures might be introduced into the design of a settlement to contain the consequences of violent disobedience, should it occur. Resorting to non-violent means of dissent is a much more likely and productive path and I next explore these options, finally suggesting that civil disobedience might be encouraged[3] by adopting a set of norms that attempts to prevent such acts as catastrophic violence while at the same time providing a socially acceptable space within which disobedience can occur.

3.2 Disobedience: The Environmental Context

The extent to which people can disobey is partly a function of the environmental context. This paradigm is particularly likely to be the case in space settlements. When environments are very extreme, groups of people are brought together under a set of life-support systems upon which all individuals depend. People depend not only on the functioning of the technology to survive, but also on the cooperation of those around them operating and servicing the technology. Thus, disobedience by one person that involves failing to carry out a required task or, in particular, violent disobedience in the form of disrupting or destroying physical infrastructure is much more likely to affect others deleteriously.

In space, life-support systems that generate interdependence between people will be pervasive in all means of existence. Liquid water for drinking must be harvested either from ice such as on the Moon, Mars or asteroids (if they have any) or brought in from an external source.[4] We do not know of anywhere else in the Solar System where water, ready to be drunk from the natural environment, exists. This, of course, is not unique to space. Most of us get our drinking water from facilities that clean and filter previously used water, so in that sense we use a life-support system. However, in space, the energies and efforts required to get access to liquid water are much greater than most places on Earth, and unlike many locations on Earth, there

(Footnote 2 continued)

total war, which often fall on a society unwilling to go outside accustomed channels.' The point about suicidal violence has a particularly powerful resonance with the extraterrestrial case.

[3]The mere idea of encouraging it may be anathema to some people, but I take the view of Zinn (2013, p. 25): 'Without those on the bottom acting out their desires for justice, as the government acts out its needs, and those with power and privilege act out theirs, the scales of democracy will be off. That is why civil disobedience is not just to be tolerated; if we are to have a truly democratic society, it is a necessity.' He goes on to say (Zinn 2013, p. 29), echoing the precepts of revolution elaborated by John Locke, 'when unjust decisions become the rule, then the government and its officials should be toppled'. The dangers of this in the extraterrestrial case, particularly when it leads to violence, I discuss in this essay. Nevertheless, the general idea that the denizens of extraterrestrial settlements should challenge their authorities and attempt to remove them when they are severely unjust seems valid.

[4]Examples of the sorts of texts that address in situ resource utilisation are Lewis (1997), Schrunk et al. (2007) and Zubrin (2012).

is no possibility of obtaining water from the atmosphere as precipitation, which acts as a buffer against a failed state or company water supply.

There is no planetary body in the Solar System, apart from Earth, that has an atmosphere breathable to people. Thus, in any space settlement, oxygen must be artificially produced, by, for example, electrolysing water or using plants to produce oxygen from photosynthesis using carbon dioxide.

In the case of food, there is no indigenous supply of readily available food anywhere else in the Solar System. Like water, this situation may not be much more extreme than for those of us who get our food from supermarkets or other shops. Few of us are self-sufficient in food production and instead rely on a vast corporate life-support system to produce our food.

The sheer extremity of space compared to Earth lies in the unrelenting requirement to artificially produce water, air and food regardless of one's location. These requirements generate an intricate interconnected labyrinth of life-support systems upon which everyone depends for their survival.[5] In essence, the instantaneously lethal conditions in outer space require a thorough and unfailing requirement for the servicing and maintenance of the complexity of machinery that enables life.

There are numerous ways in which this reality affects the political culture (Cockell 2008, 2009, 2010). It is not my intention to review the effects of the physical characteristics of the extraterrestrial environment on the nature of liberty as this has been explored previously, but some worthwhile points are to be made on the potential consequences for dissent.

First, the extremity of the space environment is likely to encourage a culture of conformity.[6] Novel political and economic arrangements that may overturn existing and well-established approaches are likely to be viewed as risky intrusions on a social system that ensures the survival of the whole population. Risk-averse attitudes are likely to propagate in the population who are concerned about their own survival and prefer stability, but also in the leaders of extraterrestrial administrations, who themselves do not want to be held accountable for decisions that result in catastrophic outcomes.[7] This pervasive culture of caution will influence dissent by making the population more suspicious of people intending to foment disobedience.

---

[5]It is in recognising this extremity that we derive the need to formulate means for civil disobedience. The problem is that the extremes of space encourage authorities to assert the idea that people must abide by the law to avoid certain destruction against the instantaneously lethal conditions. It is this tendency that civil disobedience must attempt to thwart.

[6]Cockell (2010, p. 21) states: 'The combined effects of both the physically extreme environment and the culture it propagates, and the spatial isolation of people with respect to movement and new information, contribute quite simply to the culture of conformity.'

[7]Even in very democratic extraterrestrial systems where every effort is made to create open discourse and debate, there is still the pressure on leaders to be wary of changes that threaten the populations. As Cockell (2008, p. 263) observes: 'In the confined societies of outer space, poor judgements by democratic leaders may more easily result in social disaster or death, making them more cautious, but also strengthening the power of the leaders of the oligarchy, by confirming that their experience endows them with better judgement.' In referring to the 'oligarchy', Cockell

Second, the imperative to survive in the extreme conditions of space provides authorities with the excuse to crush dissent at an early stage. Even the most benign forms of disobedience have unknown outcomes or may lead to unknown consequences. What begins as two or three people staging a sit-in in a food production unit may eventually lead to a riot resulting in damaged property. On Earth, the development of initially peaceful civil disobedience into violent rioting is dealt with by more robust police measures until the violence is quelled. The damage incurred through looting or general mayhem is then repaired. However, if the damage could result in depressurisation or destruction of infrastructure that is not easily repaired in an isolated settlement, then there is great incentive to prevent the disobedience from reaching a violent stage and therefore, as a precaution, to crush it early in its development. The authorities need not become unpopular in taking an early approach to stifle dissent. Fearful that the rebellious will threaten the lives of everyone, there is every chance that extraterrestrial settlers will demand this of their authorities and will look with approbation upon authorities that prevent violence by curtailing the activities of dissenters. For any kind of dissent to work in the extraterrestrial environment, it is likely that the protagonists will need to have considerable and widespread trust and support from the population to be able to implement civil disobedience without arousing the suspicions and antagonism of the people as well as the authorities.

Third, the period of dissent may be tolerated for a much shorter time than on Earth. Two factors are likely to make people's patience with dissent less enduring than on Earth. First, as mentioned earlier, the potential for dissent to disintegrate into violence will make people concerned. The longer dissent continues, the greater the opportunities for it to go awry. Second, even benevolent civil disobedience that disrupts services in a confined, isolated space outpost is unlikely to be sustained for long before there are demands for it to end and for a return to normal operations.

All of these factors do not align well in favour of individuals or groups wishing to disobey their authorities. All of them play directly into the hands of people who wish to find excuses to end dissent 'in the interests of the greater good'. A great challenge and problem facing the extraterrestrial society is how to judge when civil disobedience is justified and when the urge of authorities to prevent it is justified. There probably is no way to accurately make such decisions, just as political organisations on Earth find it difficult to know exactly when to intervene to end street protests. This judgement can only be made by those living on the space frontier in the specific conditions in any particular settlement. Nevertheless, recognising that there are factors that act to blunt the appetite for dissent both in the psychology of the population who demand safety and in the behaviour of administrators who are charged with providing that safety will contribute to liberty by

(Footnote 7 continued)

means the unchanging network of officials that runs a society and maintains its day-to-day function.

making people more alert to the possibilities that disagreement and disobedience are being unnecessarily curtailed.

Despite the need to be ever vigilant to the possibility that healthy dissent is being quashed or diminished, it is probably reasonable to be very fearful of violent disobedience in the space frontier.

3.3 The Problem of Violent Disobedience

The violent overthrow of government or the use of violence to dissent against particular laws is appealing. It has a certain sort of dramatic heroism about it, not helped by the fact that revolutionary leaders, however blood-soaked they may have been, have taken their place in the pantheon of historical figures. Who wouldn't want to be known forever as the leader of a violent revolution that liberated the Moon from despots?

Sometimes violent revolution may even seem just. Laudani (2013, p. 95), in quoting William Paley, observed: 'It is only when the numeric increase and growth of wealth in the colonies reached a "considerable proportion to the entire interest of the community" that the colonial revolt became "just", because, in those circumstances, "the whole happiness of the empire" was obstructed by their union.' Transplanting this to the extraterrestrial case, we can foresee instances in which space settlements, with a newfound sense of independence from Earth, could regard revolutionary struggle as the way to gain complete independence and that violence could be justified. This has been explored previously in science fiction and finds particular expression in Heinlein's (2005) *The Moon is a Harsh Mistress.*[8]

Rebellion can even be made to sound charming, as Thomas Jefferson once famously said during the 1787 Shay's rebellion: 'I like a little rebellion now and then. It is like a storm in the atmosphere.'[9]

The technological requirements for survival in the lethal conditions in space and the interdependence that results from them should immediately make it apparent why violent disobedience is an unappealing form of dissent in the space environment.[10]

---

[8]Baxter (2015a, b) has done a thorough job of investigating liberty and how it has been explored in science fiction, including instances in which there has been a resort to violent revolution.

[9]Letter to Abigail Adams, February 22, 1787. He preceded this observation with the words: 'The spirit of resistance to government is so valuable on certain occasions, that I wish it to be always kept alive. It will often be exercised when wrong, but better so than not to be exercised at all.'

[10]Rawls (1971) and others have been explicit in their view that violence is incompatible with civil disobedience. These views are generally based on the principle that violence is injurious to others and violence transgresses a basic fidelity to the law and the rule of law (even if civil disobedience involves non-violent breaking of laws, such as refusing to pay taxes). In the extraterrestrial case, the argument that violence is threatening and potentially injurious to others is augmented by the reality that violence may threaten the very survival of large numbers of people.

The notion of what constitutes 'violence' is itself worth discussing, but for the purposes of this essay, I essentially consider it to be acts that threaten people and vital survival infrastructure necessary for existence in the extraterrestrial environment.[11]

There are two sources of social pressure against violent disobedience. First, the pragmatic fact that violence may in a real way threaten the lives of everyone if it impinges on survival infrastructure.[12] Consider an individual or a group of people who decide that they will demonstrate their dislike for authorities by destroying oxygen-producing equipment or damaging the integrity of the walls of a pressurised habitat. Or merely consider a group of people who are aware of the catastrophic consequences of such violence, but much like a small riot on Earth that turns into a high-street looting frenzy, their violence under the influence of a crowd mentality results in more serious damage than originally intended. In an environment where environmentally lethal conditions bring everyone closer to death caused by technological failure, violent disobedience will always be unpopular.

This previous pressure feeds into the second source of social pressure against violent disobedience: the general culture of fear directed against violence and the desire for the peaceful status quo that ensures the safety of the settlement. This is an extreme form of Mill's (2004) tyranny of custom,[13] but it stems from a real paranoia that damage to machinery and life-support systems will kill many.

From the outset of the establishment of a settlement, there will be enormous pressure to discourage or prevent violent disobedience as a means of dissent to a much greater degree than is to be found in terrestrial environments. This discouragement has the potential to transform itself into an instrument of enforced conformity, but equally, most people, even the most libertarian and free-minded human beings, can probably agree that violence in the enclosed pressurised confines of space settlements is a bad thing.

[11]For a more developed discussion on what violent disobedience is, a good discourse is to be found in Morreall (1976).

[12]It is easy to contrive general notions of disobedience that allow violence. Zinn makes the point (2013, p. 45): 'Would not any reasonable code have to weigh the degree of violence used in any case against the importance of the issue at stake?' The problem with this idea is actually creating a calculus that allows these different issues to be weighed. And what about the case where some limited violence against the extraterrestrial infrastructure unpredictably results in a spiral into much more serious violence that destroys oxygen-producing machinery? The problem in space is that violence could kill many, if not everyone, before anyone has a chance to curtail it.

[13]Mill's idea of tyranny of custom has much in common with de Tocqueville's (1998) 'tyranny of the majority'. In extreme environments where even small errors of engineering or errant human behaviour could threaten many lives, the power of the consensus or majority view is sharpened. The tyranny of custom or the majority becomes intertwined with the very notion of survival and under these circumstances, at least to those living in the moment, the word 'tyranny' hardly seems appropriate. Indeed, as discussed elsewhere in this essay, the majority viewpoint or a custom based on prior experience may even be seen as a protector of liberty and the guarantor of survival against the lethal conditions of space.

3.4 Means to Mitigate the Threat of Violent Disobedience

Despite the foregoing, we cannot be assured that some individuals will not resort to violent disobedience. This leads to two questions: (1) how can we mitigate the desire to resort to violent disobedience? and (2) how can we mitigate the damage caused by violent disobedience if it occurs? This second question merits a separate discussion and is the subject of the following section.

In the case of the first question, there are ways to mitigate the threat of violent disobedience and indeed to minimise the general need to engage in any kind of civil disobedience. The objective of any extraterrestrial authority should be to channel disagreement and lingering resentment into productive political discussions. The types of mechanisms by which this can be done include a free press. In a small settlement, this may be nothing more than a newsletter in which opinions and points are debated.

Access to information will do much to dispel a sense of political or economic collusion and this would include information on oxygen, food and water providers, the means by which these entities are serviced and the minutes of committee meetings in which matters concerning these vital elements of a settlement are discussed. The overriding culture that underpins this access to information is that the people who run the settlement are in the same boat as the people they govern. All of them are facing an environmental foe that threatens instantaneous death. The sense of camaraderie and common purpose should therefore be enormous and should lead to a general disposition that it is in everyone's interests to discuss ideas, disagreements and points of dissent. If people feel that their ideas are being listened to, then there is less chance that they will feel compelled to engage in active civil disobedience. This culture is more easily achieved in a small settlement in which a truly participatory form of democracy where all views can be heard may be possible. The likelihood of resentment against a more distant and opaque authority becomes greater as the population size increases, mandating greater efforts to maintain active lines of discussion and debate.

Finally, the political and economic system has a bearing on the extent to which people feel engaged with the social direction in which an extraterrestrial settlement is heading. From a general point of view, political and economic systems that are more centrally run with less direction and discussion with the people being managed are likely to cause a feeling of separation between those running the settlement and those being led, with the potential for a greater number of instances in which the members of the settlement are driven to a desire for civil disobedience. Political and economic systems that are more open and allow for greater individual initiative and engagement in the management of the settlement give greater opportunity for people to express themselves and their points of view and leave them feeling less powerless and more prone to engage in disobedience. Even in the most democratic and economically open states on Earth, civil disobedience still occurs. As populations in space grow in size, there may always be small numbers of people who feel disenfranchised, however inclusive the social system. Thus, open political and economic systems cannot prevent civil disobedience or provide a failsafe way of funnelling disagreement, but nevertheless it must surely be the case that the more the people have an opportunity to express dissent openly and for their views to be heard, and acted on, the less likely it is that they will feel a need to operate *sub rosa*, to gather fellow dissenters and to engage in civil disobedience with its potentially deleterious consequences for the extraterrestrial settlement.

3.5 Engineering Safety Against Violent Disobedience

Whatever social pressures may exist to minimise violent disobedience, a settlement can never be sure that an individual or group will not resort to these measures to express dissent or that a peaceful gathering will not inadvertently degrade into violence. Therefore, engineers should plan to minimise the deleterious and potentially catastrophic consequences of violent disobedience by assuming it will happen and planning a settlement accordingly.

One could envisage several approaches to ensuring that highly destructive results of civil disobedience are contained. One approach is the modularisation of infrastructure. When habitats are all connected or when they are supplied with air, water and food by single units, there is a much greater chance that damage to that infrastructure threatens to overwhelm everyone.

Modularisation of habitats (Fig. 3.1) offers the possibility for sealing off a damaged segment of a settlement and isolating damage (and if it concerns the walls of the habitat, it potentially prevents disastrous settlement-wide depressurisation). Separation could be achieved by the provision of airlocks between segments to seal off sections of habitats. The modularisation of habitats in this way has previously been discussed in relation to advancing the general precepts of liberty. Modularisation encourages the potential for competition and choice and the development of 'districts' or 'areas' in which different ideas, cultures and approaches to efficiency may be attempted.[14] Thus, quite apart from providing a bulwark against a devastating catastrophe caused by violent disobedience, modularisation is likely to be a useful contributor to liberty within the wider settlement.

The second method of minimising damage is to ensure that vital commodities of air, water, food and power are provided by more than one production unit or

---

[14]Cockell (2010) discusses the way in which a settlement can be physically engineered to enhance liberty, including modularisation. However, one should also be mindful of the possibility that such contrivances could work in directions not foreseen. For example, modularisation of habitats might create factions and groups antagonistic to one another. These antagonisms, if they find expression in the way in which the wider settlement affairs are to be managed, could in themselves lead to civil disobedience of one group against another. So although modularisation may provide an engineered safety against violent disobedience by allowing sections of a habitat to be sealed off, it may equally be the case that modularised habitats could enhance the very disagreements that encourage violent disobedience in the first place.

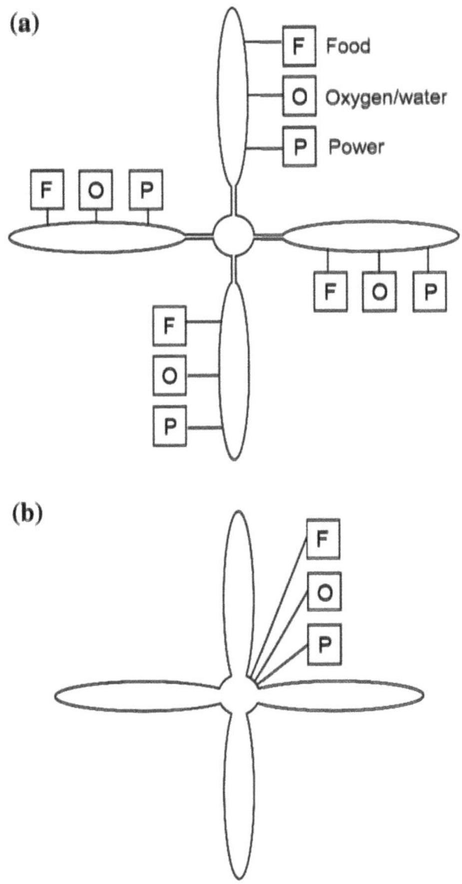

Fig. 3.1 **a** A small settlement with modularised habitats makes it easier to seal off one of these segments if violent disobedience threatens depressurisation. Also illustrated is the idea that decentralisation of major commodities [for example, food (F), oxygen (O) and power (P)] reduces the likelihood that violent destruction of one of the means of production of these commodities will threaten the entire settlement. Image **b** shows an unmodularised habitat with a single point of supply of vital commodities

machine. The destruction or damage of any given machine is then unlikely to cause catastrophic loss of vital resources for the population. The multiplication of production units could either be achieved by deliberately engineering a settlement where different areas are provided by different machines by a single provider, or where the possibility exists for multiple providers to link machines into the grid providing resources for the settlement.

As with the modularisation of habitats, the multiplication of production systems has benefits for liberty, quite apart from engineering considerations with respect to curtailing consequences of violent disobedience: the more systems of production there are, the less likely that a single authority, individual or corporation can take control of the entire settlement (particularly in vital commodities such as oxygen). Multiplication of the systems of production can enhance choice, and where these systems are operated by multiple entities they enhance competition, thus preventing the settlement from finding itself under the coercive authority of a single supplier. So again we see that by pursuing a generally liberty-seeking agenda, we may end up also providing engineering solutions to the problems that result from violent disobedience.

3.6 The Problem with a Duty of Submission to Civil Government

Paley (2002, p. 298) famously elaborated on the idea of a duty of submission to civil government: 'So long as the interest of the whole society requires it [...] so long as the established government cannot be resisted or changed without public inconvenience [...] that the established government shall be obeyed.'

Paley's view is one that in a certain sense is still implicitly agreed upon by many. Most people would prefer even a tyrannical government rather than face social collapse and anarchy caused by civil uprising against such a government. However, since Paley's time, we have also adopted different means of dissent, including civil disobedience, which allow people to act against a government in ways that do not require blind allegiance.

I bring up Paley's general idea because in the extreme conditions of outer space, it would be easy for his general precept to be adopted by a population on the basis that submission to a coercive authority is preferable to revolution, violence or any other form of disobedience that could threaten an entire population. Arguments can easily be constructed by a dictatorial authority that any form of dissent leads to unknown outcomes—including violence against infrastructure—and therefore it should be avoided at all costs. The problems inherent in the risk of violent disobedience discussed earlier in this essay give succour to the authorities who might like to endorse Paley's view as a general principle of extraterrestrial governance. The 'public inconvenience' of which Paley speaks could be interpreted as almost any form of disagreement or dissent from the level of the individual to a large group. Indeed, even a verbal assault on an extraterrestrial authority which makes some people feel uncomfortable could be contrived to be a public inconvenience.

The point to be made here is that the despotism-prone nature of the extraterrestrial environment could easily throw society back into a type of pre-Enlightenment social order where a duty of submission to an extraterrestrial authority is mandated by the need to marshal the collective support to survive in the extremities of space. These arguments can be advanced by authorities in quite subtle but nevertheless duplicitous ways. For example, the imposition of the duty to submission in order to prevent violent disobedience is a mechanism not of tyranny but of positive liberty (*sensu* Berlin[15]). By preventing dissent, the stability and safety of everyone is ensured, and within this safety net, people now have the opportunity to pursue their ambitions, which would otherwise be under threat in a

[15]Berlin (1988) provides a discussion of the separation between negative liberty (freedom from interference) and positive liberty (the capacity and freedom to do certain things). The recognition of the separation between these two types of liberty precedes Berlin (e.g. C. Bay, *The Structure of Freedom*, p. 57 and elaborated throughout M.J. Adler, *The Idea of Freedom*, Doubleday & Co., 1958). Insofar as the freedom to do a certain thing usually implies a lack of coercion against doing it (i.e. potential interference), then one can argue that there is no difference between positive and negative liberty, though Berlin's distinction is nonetheless useful.

society where there is always the possibility of violence. Thus, preventing a few miscreants from stirring up disobedience is a service to the wider liberties of the settlement. Those who seek to cause civil unrest, and thus threaten the safety of the settlement, might even, in a dark vision of Rousseau (1968; p. 64), be 'forced to be free'. The potential dissenters clearly do not understand that their actions threaten both themselves and the freedoms of others, and by quashing their plans for civil disobedience, they are not being tyrannically coerced but saved from themselves. By securing the safety of the settlement, they themselves, although they do not appreciate it, have had their liberty and life protected.[16]

In the extremities of space, safety becomes a cynical tool for manipulation and in the context of the discussion here, a method to quell civil disobedience. If an extraterrestrial settlement is not to decline into abject slavery and tyranny, then it must be forever vigilant against authorities that use the lethality of space to demand civil submission to the dictates and laws of the governing powers.

3.7 Civil Disobedience in Space

If we agree that violent disobedience is to be discouraged in space, what routes are there for disobedience? First, we might consider what we mean by civil disobedience.[17] Perhaps one of the most simple and eloquent definitions of civil disobedience was formulated by Bedau (1961, 1970), who defined civil disobedience to be thus: 'Any one commits an act of civil disobedience if and only if he acts illegally, publicly, nonviolently, and conscientiously with the intent to frustrate (one of) the laws, policies, or decisions of his government.'

[16]Rawls (1971, p. 336) observed that disobedience really only works in democracies, and using the example of authority deriving from a ruler who sees themselves with divine authority, he says, 'if the basic law is thought to reflect the order of nature and if the sovereign is held to govern by divine right as God's chosen lieutenant, then his subjects have only the right of suppliants'. The problem in the extraterrestrial case is clear enough: if the extremity of the extraterrestrial environment is held up to be an irrefragable fact of nature to which everyone must be aligned, then the excuse to manage a settlement along basically dictatorial lines will destroy any possibility of disobedience.

[17]In this essay, I take a rather broad view of civil disobedience. The reader must understand that exactly what constitutes civil disobedience (as opposed to, for example, political disobedience, radical disobedience or a variety of other definitions and nuances discussed in the literature) is a wide discussion that merits an entire volume in itself. My concern here is not to engage in a general discussion on what civil disobedience is, but instead to focus on how the settlement of space may affect general conditions for disobedience. The sort of disobedience of which I speak includes disobedience against laws, but it may also include disobedience against private institutions or the state. In the extraterrestrial environment, particularly in space settlements in the early stages of exploration, there may be very little distinction between the state, committees, private institutions, law makers, etc., thus making the sorts of discursive investigations on what exactly civil disobedience is on Earth less important in the extraterrestrial case anyway. A good volume with which to examine the nature of civil disobedience in more detail is that by Milligan (2013).

I take this to be my working definition here, where I consider civil disobedience to be about changing rules, laws and edicts of governments within some existing constitutional structure or social order and not about overthrowing the entire government in a revolution, which is a separate matter for discussion.[18] Essentially, it is about encouraging a majority to reconsider decisions or rules that it has made.[19]

Applied to the extraterrestrial environment, I take government to mean broadly 'governance', or in other words, laws, policies or decisions of any authority controlling an extraterrestrial settlement, whether that is a conventional state on Earth or a private corporation operating in space, or even a random and small group of individuals who have banded together to form an extraterrestrial settlement.[20]

In a definition of extraterrestrial civil disobedience, I explicitly include private institutions.[21] There is a likelihood that in space, settlements will be run by corporations. The result will be that private laws and profit making become intertwined[22] and private entities, although not in name, essentially become the state. Civil disobedience in these circumstances is therefore most definitely directed against a private institution.

Bedau's view is consistent, if not in detail but in broad scope, with the non-violent philosophies of civil disobedience offered by Ghandi (1961) and Thoreau (2010). The latter's view is perhaps known through this quote (2010, p. 17):

---

[18]Bedau's definition is broadly consistent with others who also see civil disobedience to be about changing the behaviour of existing government decisions. Rawls (1971, p. 364) for example, defines civil disobedience as 'a public, nonviolent, conscientious yet political act contrary to law usually done with the aim of bringing about a change in the law or policies of the government.' Zinn defines it as 'the deliberate, discriminate violation of law for a vital social purpose'.

[19]Singer (1973) elaborates on this view of disobedience and sees disobedience as a strength rather than a weakness of democracies since it provides a means for a minority to change what on the face of it are legitimate decisions taken by a majority through democratic procedures. The problem, as Singer recognises, is that majorities can make wrong decisions on things that do not matter to them, but are vitally important for a minority. Disobedience provides the means for a minority to register this strength of feeling.

[20]In examining the civil disobedience of Thoreau, Milligan (2013, p. 65) reminds us that 'the state is nothing more than a political body which helps to get some important things done. Accordingly, we should not, out of any sense of loyalty to the state, accept instructions to engage in acts of injustice towards others.'

[21]In some terrestrial definitions, private institutions have been excluded, for example by Raz (1979, p. 264), but I suggest here that the extraterrestrial environment, which is likely to be opened up vigorously by private concerns who will exclusively run some outposts, has a particular possibility of generating privately enforced despotism and therefore civil disobedience against such institutions may be necessary.

[22]Persson (2015, p. 133) explores this entire problem and states that 'Given the overarching aim for the corporation of making money for the shareholders, "free thinkers" and dissidents will not be of any interest to the company.' He elaborates on the basic problem thus (Persson 2015, p. 132): 'If the organisation that governs the settlement is the same as the corporation that supplies the life supporting services, it will create an extreme asymmetry in power between the rulers and the ruled that will make it very difficult for the settlers to claim any respect for their civil liberties.' He suggests that there needs to be robust separation between those making the laws and those running the settlement.

'but if it is of such a nature that it requires you to be the agent of injustice to another, then, I say, break the law. Let your life be a counter friction to stop the machine.'[23]

Ghandi saw civil disobedience as a mechanism for bending the government to the will of the people and a pressure valve for preventing worse resentment from eventually erupting with potentially more violent outcomes.[24] At the core of this view of civil disobedience is the notion that breaking the law may be justified if injustice or despotism is to be held at bay.[25] In the extraterrestrial environment, the crucial question is whether the lawbreaking ultimately threatens the supply of some vital commodity such as oxygen. When it does, then it may unleash violent disobedience from other parties (which one wishes to avoid) or may soon lose its support from the community at large who fear the loss of such a commodity.

Breaking the law may not be required to slow or reverse decisions taken by an extraterrestrial settlement authority. Mere disruption to administrative or bureaucratic services may be sufficient to bring a counterweight against laws and regulations that are unpopular. The possibility of law-abiding disruption and lawbreaking in more extreme cases opens the field for a variety of colours of civil disobedience.

The types of civil disobedience one might imagine in an extraterrestrial settlement could be varied, but let me attempt to suggest some possible ideas: (1) sit-ins in habitats and spaces necessary for the administration of the settlement in order to cause bureaucratic disruption; (2) protests that cause disruption to management; (3) refusal to pay taxes or similar financial obligations; (4) disruption to production of important, but not immediately critical, services such as food and non-vital domestic commodities.[26]

As for violent disobedience, the objective of an extraterrestrial settlement must be to minimise the compulsion that people have to engage in disobedience in the first place. A free press and open access to information are two vital prongs of such an effort. With regard to non-violent disobedience, there are other facets of an

---

[23]Thoreau (2010, p. 17) goes on to say, 'What I have to do is to see, at any rate, that I do not lend myself to the wrong which I condemn.'

[24]Ghandi's point of view is explored by Haksar (1986) and Milligan (2013).

[25]There is much disagreement on whether breaking the law is ever justified since it may foster a disregard for all laws. Certainly a general culture of lawbreaking is likely to erode social order, but as Zinn (2013, p. 12) points out, '… this is like arguing that children should be made to eat rotten fruit along with the good, lest they get the idea all fruit should be thrown away. Isn't it likely that someone forced to eat the rotten fruit may because of that develop a distaste for all fruit.' As he states earlier in his treatise (p. 19), 'If the effect of civil disobedience is to break down in the public's mind the totalitarian notion that laws are absolutely and always to be obeyed, then this is healthy for the growth of democracy.' This may be particularly pertinent to the extraterrestrial case since extraterrestrial environments are prone to tyranny.

[26]This is a non-exhaustive list and the specifics of the opportunities offered to engage in civil disobedience will depend on how a settlement is run and what sorts of bureaucratic machinery and administration can be disrupted for the purpose of civil disobedience. This can ultimately only be known by those living on the frontier.

extraterrestrial settlement that can lend a helping hand in allowing dissent but preventing it from degenerating into destructive tendencies.

There must be places to organise dissent and places where dissent can take place. This amounts to habitat space where people can physically meet, debate and organise. Ancillary to this is the requirement that there is not the opportunity for settlement authorities to incessantly monitor conversations and meetings to determine from where dissent might be emerging, thus empowering them with the information to crush it. In an extraterrestrial settlement, cameras and other monitoring systems provide an excuse, and in some cases an engineering requirement, for continuous monitoring. The surveillance state[27] that emerges from an engineering need also provides the infrastructure to monitor dissent and with it, the apparatus of safety can again be turned against the people. Thus, there needs to be a surveillance system that is minimal in its scope and directed solely towards engineering, and the information on how this system is used and monitored needs to be open and exposed to free discussion and discourse. What the above suggestions amount to is nothing more controversial than the idea of free and open spaces where people can gather, debate and dissent without draconian oversight.

In confined locations in outer space, an additional question is how much civil disobedience can be allowed. It is probably unrealistic to think that a large number of disparate groups would all decide to be disobedient at once, but then again, a general feeling of discontent that leads to two or three groups of people carrying out civil disobedience could collectively amount to a serious disruption of vital services. It seems that the solution to this problem is again to ensure that there are spaces and forums for open discussion so that to some degree a consensus can emerge on the 'allowable' amount of dissent at any given time.[28]

We might conclude by noting that the abovementioned ideas are by no means exhaustive. Laudani (2013, p. 119), quoting Howard Zinn, explains: 'Civil disobedience should thus be continually "enriched" with countless possibilities and tactics not yet imagined that are able to control and focus the disorder of civil disobedience toward the creation of a life more human.' Nowhere is this more the case than outer space. As we do not know from exactly where the greatest threats to liberty come from in space, at least in detail, and we do not know the details of

---

[27]Cockell (2010, p. 17) discusses the problem of the surveillance state in extraterrestrial settlements and later (Cockell 2015) elaborates on how the tendency for safety to degrade into an all-embracing culture of monitoring might be deliberately avoided when establishing extraterrestrial settlements.

[28]Assuming that there is a collective sense of responsibility to the settlement, it is not unrealistic that people would choose to self-regulate the extent to which they carry out disobedience. Rawls (1971, p. 374), in recognising that there is an upper limit of the 'ability of the public forum to handle such forms of dissent', avers that 'The ideal solution from a theoretical point of view calls for a cooperative political alliance of the minorities to regulate the overall level of dissent [...] what seems called for is a political understanding among the minorities suffering from injustice.' These ideas seem idealistic, but if faced with instantaneous death from depressurisation and extreme isolation, it may not be unrealistic to expect the denizens of extraterrestrial settlements to cooperatively manage and exercise restraint in their use of disobedience.

engineering, political and economic systems that might be used to prevent dissent, then it must be the case that those on the space frontier should always be engaging in discussion about how to allow for open dissent and even disobedience within the confines of isolated settlements.

3.8 Disobedience and a Potential Problem with Strike Action

The potential problems with implementing civil disobedience may find expression in strike action. In particular, the problem of employee–employer conflict is a matter of importance, particularly in crucial industries such as oxygen production and their allied industries that produce spare parts and raw materials. The closure of these industries could threaten a settlement, and thus coercive efforts to keep them open encourage economic tyranny from the authorities who wish to ensure that they either protect a profit motive or if they are a government-run entity (non-profit), they are at least being seen to implement their responsibilities in the efficient and safe operation of the settlement.

Figure 3.2 shows the structure of an extraterrestrial strike, which I derive from the terrestrial case discussed by Crouch (1982). The graph shows the price (in arbitrary units) that an employer and employees are willing to pay to end a dispute, against time. For the terrestrial case, curve A depicts the cost to the employer over time after the start of a strike. After an initial period after the strike starts, during which an employer makes a calculation to stand against the strike and their resolve strengthens (the price they are willing to pay to end the dispute reduces), the cost to an employer increases over time as greater profits are lost due to the strike action or the implications (in the case of a government-run facility) for failure to return operations to normal become more drastic. The price they are willing to pay to settle the dispute therefore increases over time. By contrast, the employees, initially emboldened by their strike action (and they increase their price for settlement as they dig in), soon find their price for agreement reduces as prolonged strike action denies them their livelihood (B). The employees and employer will reach an

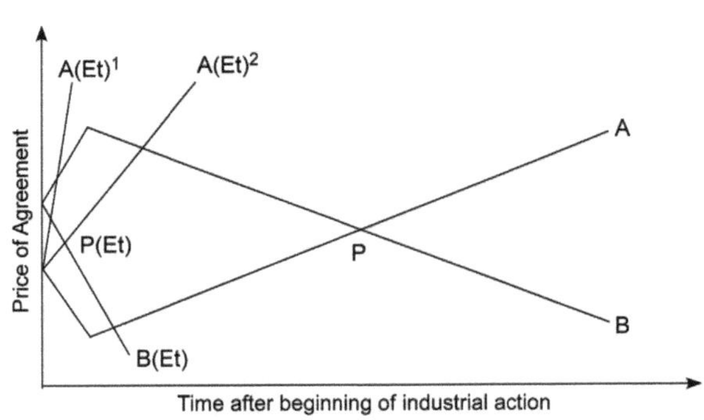

Fig. 3.2 Economics of extraterrestrial strike action. In many extraterrestrial industries (except the non-vital), strike action could be short-lived and unpopular (see text for detail)

agreement where lines A and B intersect (point P) at some time after the beginning of the strike.

Crouch increases the complexity of the relationship between the curves as he relates them to the growing loss of profits in the entity concerned. Nevertheless, the basic curves described are sufficient to illustrate key points about the extraterrestrial case. First, the price at which employees will settle is likely to be, in the extraterrestrial case, lower than the terrestrial case [B(Et)] much earlier on, particularly for vital industries. This is because the cost to the employees of the shutdown of production of a commodity required in the extreme environment of outer space is likely to be high. It is likely to encourage much greater unpopularity among other people dependent on the products of those industries, increasing the peer pressure to withdraw from such behaviour. The price of agreement for a corporation or the organisation producing the commodity is also likely to increase more quickly than in the terrestrial case [$A(Et)^2$]. This is due to the employer's concern that productivity shutdown will result in loss of confidence in their capacity to produce vital commodities and will become a possible threat to the settlement as a whole. The price of agreement for the extraterrestrial authorities that oversee the entire settlement will be even greater [$A(Et)^1$] because they are likely to be held responsible for allowing production from a vital industry to be curtailed and will be under great pressure to bring the action to an end. Thus, extraterrestrial strikes in vital industries may be short-lived [P(Et)].

Given these conditions, it may be unlikely that any vital industry would experience strikes because the cost of the agreement is so great and the effect of a strike so unpopular that an incentive exists for employees and employers to prevent disputes reaching the stage of a strike in the first place and to reach early agreement. Except in the most trivial non-critical industries, strikes will hardly be a viable mechanism for employees to vent grievances against their employers in most extraterrestrial environments, except in the extreme case where the employees have suicidal tendencies. This would seem to considerably reduce employee power in the workplace in the event of a serious grievance against their management. It may even be the case that such strikes would be outlawed.

However, one should be mindful of another possible effect. It could be the case that the mere threat to strike would have a much greater influence on the employer than would be the case on Earth, and so in this sense the employees of vital industries could become more powerful. This possibility will depend on the employer's assessment of whether the threat of a strike could realistically be implemented by the employees, the type of industry affected and the legal status of strike action.

Whatever the direction strike action takes, the foregoing serves to illustrate how economic principles will govern some types of disobedience in space, particularly strike action. Strike action also illustrates a more general point that the extremity of the space environment yet again can be seen to provide authorities with an excuse to limit civil disobedience. As strike action could threaten the provision of the most basic commodities upon which human lives depend, then authorities may seek to stop it to prevent disastrous loss of production (and to save their own credibility), and inhabitants of an extraterrestrial settlement may seek to stop it because of their fear of the potential consequences. Thus, the people and the authorities collude to prevent disobedience.

3.9 Formalising Channels for Extraterrestrial Disobedience

Are there ways in which civil disobedience might be formalised in laws and edicts within a space settlement that allow for dissent, but within a framework that prevents catastrophic violent disobedience? Some people might say that establishing rules or boundary conditions for civil disobedience is antithetical to the very idea of civil disobedience, which is to disobey rules. However, there are likely to be some actions and behaviours that most people can agree are potentially disastrous for a space settlement and which could be used to circumscribe, within laws, a protected realm within which disobedience can be accomplished. For example, we can probably all agree that civil disobedience that results in damage to infrastructure is something we wish to avoid.

On Earth, attempts to define a set of behaviours that could be considered 'universal' norms have focused on the idea of 'civility'. The claim revolves around the idea that there are aspects of civil conduct that everyone can agree on, even those about to engage in disobedience. Milligan (2013) lists a set of these, which I have reproduced below. The only one I have altered is the fifth one in which I have included 'violence against survival infrastructure' as a particular extraterrestrial problem (although not exclusively as destruction of vital infrastructure in isolated polar stations, submarines and other small enclosed settlements on Earth can be lethal).

The behaviours listed below might allow one to permit disobedience without threatening the survival of the population[29]:

- Respect for others and recognition that other humans are fellow humans;
- Rejection of hate speech;
- Avoidance of acts driven by hatred;
- Try to avoid violence;

[29]Milligan's set of civility norms (Milligan 2013, p. 36) follows a tradition that includes the concepts of civility advanced by Gandhi. It broadly encompasses concerns such as good manners and polite demeanour. Although these concepts are notoriously difficult to pin down, they nevertheless provide some sort of cultural brake on excessive violence, and in the extraterrestrial case at least caution people to be ever mindful of the potential disastrous effects of violence unleashed on a small enclosed population in a lethal environment. Any such mechanisms that provide checkpoints on people's behaviour, even if only vague cultural norms, must surely be a good thing.

- Always avoid violence against survival infrastructure and reckless endangerment of others;
- Avoidance of cruelty.

The notion of introducing defined sets of norms to allow for a legal and behavioural space within which civil disobedience can occur cannot offer a foolproof way of directing civil disobedience towards positive ends. First, there can never be certainty that people will conform to these norms, and while that uncertainty exists, enormous peer pressure and latent fear will always exist within an enclosed extraterrestrial settlement against anyone considering civil disobedience. Public pressure to conform will still exist. Second, the elaboration of such rules can itself be used as an instrument against people. Consider for example 'avoidance of cruelty'. A group of dissenters plans to organise a strike in a food production unit. They plan no violence and their motives make it abundantly clear that they will do nothing that endangers the lives of anyone in the settlement. However, by staging a strike in the food production unit, they deny some domestic animals some food for a few days. This violates the avoidance of cruelty norm and the general and even legal disapprobation of the settlement can be used to quash their dissent.

The point is that the so-called 'norms' cannot be accurately circumscribed and arguments can be constructed to demonstrate that a norm is being violated when authorities or the people within a settlement wish to put an end to a particular plan of dissent. Consequently, the norms probably should not be instantiated in law as they will become instruments of tyranny. They should instead be a stated set of ideals to which space settlers subscribe in the implementation of disobedience, a form of 'ten commandments' whose moral force is brought about by the fact that ignoring them or violating them is likely to result in catastrophic death in the settlement. They are, in essence, a code of extraterrestrial civility that recognises the need for collective responsibility in space, giving reassurance to those who would oppose disobedience at all costs while at the same time overtly showing that civil disobedience is allowable.

3.10 Conclusion

Disobedience, as on Earth, is one of the most important channels through which people will dissent in space. In the extreme and lethal environment of outer space, violent disobedience becomes a significant challenge to the community. Political methods must be used to discourage it, but engineering approaches must be used to provide the means to isolate and seal off areas of settlements where violent disobedience threatens the lives of many. Non-violent civil disobedience is a plausible channel through which individuals can remove or change unwanted governance or laws. A set of norms might be used to discourage violent disobedience and thereby make non-violent disobedience more socially acceptable.

References

Baxter, S. (2015a). The cold equations: Extraterrestrial liberty in science fiction. In C. S. Cockell (Ed.), *The meaning of liberty beyond earth* (pp. 13–31). Berlin: Springer.
Baxter, S. (2015b). The birth of a new republic: Depictions of the governance of a free moon in science fiction. In C. S. Cockell (Ed.), *Human governance beyond earth: Implications for freedom* (pp. 63–79). Berlin: Springer.
Bedau, H. (1961). On civil disobedience. *Journal of Philosophy, 58*, 661.
Bedau, H. (1970). Civil disobedience and personal responsibility for injustice. *The Monist, 54*, 517–535.
Berlin, I. (1988). Two concepts of liberty. In I. Berlin (Ed.), *Four essays on liberty* (pp. 118–172). Oxford: Oxford University Press.
Cockell, C. S. (2008). An essay on extraterrestrial liberty. *Journal of the British Interplanetary Society, 61*, 255–275.
Cockell, C. S. (2009). Liberty and the limits to the extraterrestrial state. *Journal of the British Interplanetary Society, 62*, 139–157.
Cockell, C. S. (2010). Essay on the causes and consequences of extraterrestrial tyranny. *Journal of the British Interplanetary Society, 63*, 15–37.
Cockell, C. S. (2015). Extraterrestrial liberty: Can it be planned? In C. S. Cockell (Ed.), *Human governance beyond earth: Implications for freedom* (pp. 23–42). Berlin: Springer.
Crouch, C. (1982). *The logic of collective action*. London: Fontana.
de Tocqueville, A. (1998). *Democracy in America*. England: Wordsworth.
Ghandi, M. (1961). *Nonviolent resistance*. New York: Schocken Books.
Haksar, V. (1986). *Civil disobedience: Threats and offers: Ghandi and Rawls*. Delhi: Oxford University Press.
Heinlein, R. A. (2005). *The moon is a harsh mistress*. London: Hodder Paperbacks.
Laudani, R. (2013). *Disobedience in Western political thought: A genealogy*. Cambridge: Cambridge University Press.
Lewis, J. (1997). *Mining the sky: Untold riches from the asteroids, comets and planets*. New York: Basic Books.
Locke, J. (1988). *Two treatises of government*. Cambridge: Cambridge University Press.
Mill, J. S. (2004). *On liberty and other writings*. Cambridge: Cambridge University Press.
Milligan, T. (2013). *Civil disobedience: Protest, justification, and the law*. New York: Bloomsbury.
Morreall, J. (1976). The justifiability of violent civil disobedience. *Canadian Journal of Philosophy, 6*, 35–47.
Paley, W. (2002). *The principles of moral and political philosophy*. Indianapolis: Liberty Fund.
Persson, E. (2015). Citizens of Mars Ltd. In C. S. Cockell (Ed.), *Human governance beyond earth: Implications for freedom* (pp. 121–137). Berlin: Springer.
Rawls, J. (1971). *A theory of justice*. Cambridge, MA: Harvard University Press.
Raz, J. (1979). *The authority of law: Essays on law and morality*. Oxford: Clarendon Press.
Rousseau, J.-J. (1968). *The social contract*. Harmondsworth: Penguin.
Schrunk, D. G., Sharpe, B. L., & Cooper, B. L. (2007). *The moon: Resources, future development, and settlement: resources, future development and colonization*. Berlin: Springer.
Singer, P. (1973). *Democracy and disobedience*. Oxford: Oxford University Press.
Thoreau, H. D. (2010). *Civil disobedience*. Seattle: Pacific Publishing Studio.
Zinn, H. (2013). *Disobedience and democracy: Nine fallacies on law and order*. Chicago: Haymarket Books.
Zubrin, R. (2012). *The case for mars. The plan to settle the red planet and why we must*. New York: Simon and Schuster.

Chapter 4
Lunar Labor Relations

James S.J. Schwartz

Abstract This chapter defends the claim that at least some essential service strikes will be morally permissible in lunar societies. I argue by analogy with essential service strikes in medicine. The duty to provide medical care is not borne exclusively by nurses and physicians but rather is borne by a wider healthcare community, which includes hospital administrators, insurance companies, etc. If it is permissible for hospital administrators and insurance companies to compete for their interests by making decisions which might negatively affect patient care, then nurses and physicians should have at least a limited right to bargain for their interests (and the interests of their patients), which should include, in extreme circumstances, a right to withhold their labor. Similarly, lunar essential service workers, such as low-level employees in the air production industry, ought to retain the right to strike. The duty to provide breathable air is borne by a large "air production" community that includes air production staff, administrators, and corporations, as well as regulators and the voting public. If other members of this community were to make decisions which resulted in especially egregious working conditions for air production staff, then these employees would have the right to strike to improve their working conditions.

Keywords Business ethics · Labor relations · Essential services · Healthcare ethics · Nursing ethics · Bioethics

4.1 Trouble in Peary Crater City

Those oxygen crackers get all of the glory. Adverts with sparkling white pressure suit clad "heroes" delivering boxes of air to grateful families. (In reality the regolith gets every-where *and crackers are some of the dirtiest scum employed by Lunar Oxygen Systems Technology Inc.—yeah, nobody noticed the unfortunate acronym until it was too late.*

J.S.J. Schwartz (✉)
Department of Philosophy, Wichita State University, 1845 N Fairmount, Wichita, KS 67206, USA
e-mail: james.schwartz@wichita.edu

C.S. Cockell (ed.), *Dissent, Revolution and Liberty Beyond Earth*,
Space and Society, DOI 10.1007/978-3-319-29349-3_4

I hear your job is forfeit if you're caught saying you work at LOST Inc. in space!) The royal treatment, they get. Saviors of Peary Crater City, they are, putting their lives at risk performing the vital work of running the volatile extraction machinery out on the surface. Always get to drink on the house. Always get all the dates. Can't even compare them to us under-appreciated, over-worked, and under-paid CO_2 *scrubber techs. (Another unfortunate acronym, since all corporate seems to think we are is just another "cost" they need to minimize...) We're the human equivalent of trees, but there aren't any nature lovers around.*

Don't get me wrong. The work we do is important-all 12 hours a day of it. It's our job to make sure that each module's set of carbon dioxide scrubbers is in good working order, and there are lots and lots of scrubbers to monitor, repair, and replace. Peary Crater law dictates that every sealed environment contain a minimum of two scrubbers which must be subject to regular "professional" inspection and maintenance. And of course the Peary Crater legislators seem to think just about every door in the city should be replaced with an air-seal, even the doors to folks' toilets. It's good for the door manufacturers and it's good for LOST Inc., I guess. And I suppose it helps guard against the worst depressurization might bring your way. But anyway, where was I?

Oh, yeah, the working conditions. Your average scrub job takes anywhere from 15 to 120 minutes. 15 if there aren't any problems; 120 for a complete replacement. Trouble is they hand us jobs under the assumption that they'll average out at 30 minutes, but lately more and more wind up taking 120, and they dock our pay if we have to cancel a job. And when a job goes long we're late to the next, and when folks have to wait they get mad, and when folks get mad they sound off at us. That really wears on you. To hear how much of a loser dimwit waste of resources you are; how awful it is you've got dirty hands; why you're not more like the crackers in pristine suits they see on the adverts. Those cable repairmen back on Earth have nothing on us!

And recently it's gotten worse. At our employee meeting the other week we learned that, due to increased competition and tightened regulation, LOST Inc. was going to have to make a 15 % staff reduction. And, you guessed it, the trees were going to bear the brunt of this burden. We were looking at a 10 % increase in our daily workloads; another 10 % of jobs we probably couldn't make; which translates to a nice 10 % pay reduction when you figure all of the extra docked pay (hey—twice the savings for corporate!). How much angrier could folks be with us? We haven't found out yet. We could barely work in the conditions before-and even then our schedules pressured us into making too many mistakes, which means one visit could turn into two or four-but these new demands are just preposterous. And so we took a vote and decided to go on strike, right then and there. No way we could let LOST Inc. wring any more out of us. So until they not just rescind the threat of reductions but promise to increase our numbers, increase our pay, and reduce our workloads, we aren't touching a damn scrubber.

Now, there aren't enough trained techs around to make up the work—hell, the trees at other companies are having the same kinds of problems and might even join in on the fun. Meanwhile, some areas of Peary Crater City are approaching dangerous concentrations of CO_2*. Under pressure from the governor, LOST Inc.'s response has been to increase oxygen production and pump more fresh air into the system. But that's a temporary and expensive measure. The crackers are starting to complain about the extra hours, and I've heard rumors that they might start a work-to-rule strike, or even a work stoppage. I bet that would get corporate's attention. But it in the meantime we'll still be sitting around not fixing the scrubbers. We're all hoping corporate will cave well before the air gets too poisonous. I'm starting to wonder whether we'll cave first? The few of us with families made sure they knew where the best maintained air systems were, but we certainly don't want blood on our hands. I sure hope no one dies...*

What should our ethical analysis be of these workers' actions? By withholding their labor they are interfering with the public's access to clean, breathable air—a resource that the public requires *continuous* access to. If any sort of service is essential for the well-being of Peary Crater City, it is air production and distribution. Thus if there is any employment sector which has the greatest obligation to refrain from work stoppages, it is the air-production and distribution sector. Such work stoppages could place the entire population at risk of asphyxiation and death. *Prima facie*, air producers are morally obliged to refrain from striking, and air production workers must place the needs of the community above their own narrow self-interest.

But are matters necessarily so simple? To say that the essentiality of air production so clearly places these obligations on air industry employees could ultimately require us to accept that these workers have an obligation to provide their labor *regardless* of their working conditions. This would turn air industry employment into the moral equivalent of slavery. Essential though the services they provide may be, we should not tolerate a situation in which employees of the air industry have no choice but to work in extremely hostile conditions.

We thus face a dilemma: It would be morally unacceptable for air production to be halted, even for a short period of time. But it would also be morally unacceptable to prohibit air production employees from exercising various forms of protest, including strikes, in so far as these activities might be the only effective measures for mitigating hostile working conditions. Is there any way to escape this dilemma? Or will the restriction of worker's rights be a necessary component of essential service industries in space?

These questions are especially vexing in the sense that we have no closely analogous terrestrial cases to use as models. There is no terrestrial industry which is as vital to the persistence of society as the air production industry is to the persistence of Peary Crater City. On Earth there is no *manufactured* resource on which we absolutely depend for survival. Even the most "essential" resources we can do without for longer than seconds-medicine, water, food, etc. But not so with air, which we require on a second-by-second basis. In lunar societies like Peary Crater City air is a resource that must be manufactured, distributed, and maintained. As such we can make no direct analogies between extant discussions of essential service strikes and lunar air production strikes.

However, we should not dismiss existing philosophical and ethical discussions as irrelevant to the issue at hand. We can at least say the following: If essential service strikes are *impermissible* in terrestrial settings, then *a fortiori* essential service strikes *remain impermissible* in extraterrestrial settings (since they are of all the more 'essence' in space). On the other hand, if, when salient conditions have been met, at least some terrestrial essential service strikes are permissible, then *mutatis mutandis* it may be possible to provide a structurally similar justification for the permissibility of some extraterrestrial essential service strikes.

The position I defend is one according to which at least some essential service strikes are permissible, in both terrestrial and extraterrestrial settings. To do so I borrow from recent arguments for the permissibility of nurse and physician strikes.

As I discuss in Sect. 4.2, nurses and physicians are not exclusive bearers of the obligation to provide healthcare—this obligation is also borne by hospital administrators, insurance companies, politicians, and society as a whole. All members of the wider healthcare community make decisions that impact the quality and availability of healthcare, and the community as a whole has an obligation to arrange itself in a way that minimizes the injustices suffered by its members—including nurses and physicians. In certain situations, strike action by nurses and physicians may be the only way to raise awareness of and begin to rectify the injustices related to working conditions and patient care initiated by other members of the wider healthcare community.

After establishing the permissibility of nurse and physician strikes I return in Sect. 4.3 to the issue of extraterrestrial essential service strikes, and in particular to the hypothetical case outlined above. I argue that, as in the case of the duty to provide healthcare, the duty to provide adequate breathable air is not borne exclusively by air production employees but by lunar society as a whole, including air production companies, legislators, and the voting public. This wider "air production community" is in turn responsible for the working conditions of employees such as LOST Inc.'s CO_2 scrubber technicians, and is furthermore responsible for rectifying any injustices that these employees might suffer in the pursuit of, e.g., cheaper air. Strikes are a permissible way of calling attention to the injustices air production staff face, provided that absolutely vital systems are not compromised. Following a comparison of Frederic Young's discussion of solar energy station worker strikes in Sect. 4.4 I make a brief conclusion in Sect. 4.5.

4.2 The Ethics of Healthcare Strikes

There is a strong moral presumption against disrupting essential services, such as medicine, utilities, and law enforcement. The disruption of any of these services to a large enough population for a long enough time could lead to great harms, such as injury and death. It is difficult to see how the typical intended results of strike action —the improvement of the working conditions and pay of the strikers—could make up for the quite palpable harms such actions could produce. Thus there are utilitarian grounds for prohibiting strike action in the essential services. There are also Kantian grounds—strikes use the public as a mere means to improving the lives of strikers. In fact, the "success" of such strikes depends entirely on creating a sufficiently perturbed public.

It might seem, then, that there is nothing that could morally recommend strike action in the essential services. However, this judgment would be too hasty. There are, I argue, situations in which we might condone strike action in the essential services. How is this case to be made?

One initially promising strategy is to question the coherence or the usefulness of the concept of "essential service." If there is no principled way of determining whether a service is essential, then classifying a line of work as an *essential* service cannot in itself be of moral significance, and thus cannot assist in a judgment about the permissibility of essential service strikes. For instance, Marley Weiss argues that in the present-day United States, the notion of an essential service is not practically useful because almost nothing qualifies as a truly essential service:

> …almost no service can properly be called "essential". Nearly always, there is a reasonable substitute… It is precisely because a strike normally occurs at only one employer, or one facility, that consumers normally retain many options. (Weiss 2000, p. 195)

Nevertheless there are situations in which consumers' options are incredibly limited, such as when an employer has a monopoly over a service in a particular geographical area. For instance, if a city has only one hospital and that hospital's nursing staff strikes, then that city's medical care will be severely comprised. Moreover, even if it were in principle possible to seek help elsewhere—if there was a second hospital, for example—it would still likely be that many patients' circumstances—their insurance or their medical condition, for example—would prevent them from seeking help elsewhere. Substitutes are not always available in practice, especially for the poor. Weiss's point is of course a contingent one; I would expect that she would agree that the notion of an essential service becomes useful when consumers' options are extremely limited.

There is nevertheless something odd about describing healthcare as an essential service. Clearly not all medical services are essential—at least in the sense that someone's life will be cut dramatically short if care is not provided in the very near future. Many operations and treatments can be postponed without causing significant harms to patients. Thus, Susan Braithwaite is somewhat skeptical that much sense can be made of what counts as an essential medical service:

> As a result of different cultural backgrounds and personal value systems, each layperson or physician likely would give a different answer. It is difficult to agree on a definition of medical need or essential services. (Braithwaite 2000, p. 477)[1]

Gerald Dworkin is similarly skeptical about the circumscription of essential medical care:

> …the attempt to restrict treatment to emergency cases only is in itself a breach of the basic professional ethic, and the attempt to define an emergency case in a rational and caring way almost impossible. (Dworkin 1977, p. 79)

I suspect the concerns here arise from the following kind of reasoning: It is perfectly sensible to say that medical attention is essential when someone's life, or quality of life, is at stake. But this does not tell us how immediate the risk of life must be for the attention to be essential; nor does this tell us how compromised someone's quality of life must be before attention becomes essential. A person who will die within hours without care is clearly in essential need of medical care. As is someone who will without treatment lose the ability to eat and drink on their own.

---

[1]It should be noted that Braithwaite's point is presented as an objection to physician strikes—based on the potential for physicians to *mis—classify* essential care as non-essential care.

But what if the person could be saved so long as they are treated anytime over the next week or month? Is it a dereliction of duty to withhold treatment up until the time that care becomes an immediate need? If not then how immediate must the need be before it becomes immoral to withhold or delay care? What if the person will only loose one lung if they go without care? Does this constitute a situation in which it is essential to receive immediate care? If not, then how compromised must one's quality of life be before one requires immediate care? I suspect these questions do not have decisive answers.

A conservative reaction is therefore to prohibit all or nearly all healthcare strikes, since it is always possible that during a strike some emergency could go unaddressed (this is the position that Dworkin opts for). Only I do not think this is the most sensible reaction. Though medical care may not divide neatly into emergency and non-emergency situations, and consequently, medical services into essential and non-essential, we can at least conceive of essentiality as a continuum. As Cristian Muñoz relates, "we can say that the essentiality of a given service is determined by the relationship between the level of harm that can likely be produced when the provision of that service is interrupted and the time of stoppage" (Muñoz 2014, p. 652). We can in turn say that medical strikes become increasingly difficult to justify as the strike's anticipated level of harm increases. Thus, emergency room staff might never be justified in striking, but dermatologists and optometrists might easily justify a strike (and might do so on behalf of more "essential" hospital staff). Nevertheless this fails to tell us whether the justificatory burden for strike action even in the least essential medical services can be met.

If weaseling away at the notion of essential service fails to decide matters, then on what other grounds could the permissibility of healthcare strikes be demonstrated? I think the most attractive strategy begins by recognizing that medical care is not a simple binary relationship between provider and patient but rather is a much more complex affair involving, in addition to patients and providers: hospital administration, insurance providers, governments, and voting citizens. In other words, what needs to be recognized is that the obligation to provide adequate medical care is not solely borne by workers in direct contact with patients but is also borne by the entire healthcare system. This means that, when the system fails to provide adequate medical care for patients, or forces healthcare workers to work in deplorable conditions, any number of agents could be responsible. Nurses and physicians could fail to exercise due diligence, perhaps. But hospital administrators could fail to hire sufficient staff to cover patient needs. Insurance companies could deny coverage making subsequent care unaffordable to the patient. And government officials, perhaps acting on the will of the electorate, could fail to provide the under-insured with adequate means for seeking healthcare.

Suppose at some hospital there is an alarming shortage of healthcare workers, where nurses and doctors routinely work extended hours and cannot provide each patient with the level of care that they believe, in their professional opinions, is appropriate. But the hospital could only reduce working hours and improve patient care by hiring more staff—something the administration denies is fiscally possible.

In such a situation a nurse or physician strike may be the only way to place sufficient pressure on administration to bring about improved working conditions, and ultimately, improved patient care.

It is in this kind of situation that I think it *could* be morally permissible for the nurses and doctors to strike. This is because the strike is likely the only way of rectifying the occurrent injustices: hostile working conditions and inadequate patient care. The strike may temporarily severely compromise patient care, but if it is successful it will in the long run lead to much better working conditions and much improved patient care. To say that the likely harms caused by the strike present an overwhelming case for its impermissibility would imply that, in this case, the suffering of nurses and physicians is a "precondition for meeting the needs or satisfying the interests of others," and I agree with Peter Railton that this is a system that "we can no longer tolerate" (Railton 1980, pp. 201–202).

The most cogent elucidation of this strategy comes from Paul Neiman (2011).[2] Though Neiman's interest is exclusively with nurse strikes, I believe that what he says applies with little modification to physician strikes. As Neiman argues, the quality of healthcare is a function of the overall healthcare system, and is not uniquely determined by practicing nurses (and physicians). There are push and pull factors throughout the system that can positively and negatively affect patient care. Thus, Neiman characterizes the healthcare system as comprised of a variety of "competing interests":

> Hospitals, healthcare workers, insurance companies, taxpayers, and government all work together to provide healthcare for the people of the community. But each of these groups also has interests in conflict with this goal. Insurance companies seek to increase profits, and the actions they take to do so negatively impact the quality and availability of healthcare. Hospitals may seek to make more money, or to keep costs down, and doing so may negatively impact the quality and availability of healthcare. Politicians may seek to please taxpayers by reducing or increasing government programs that may negatively or positively impact the quality of healthcare. It is in this context that the morality of nurses' strikes ought to be considered. (Neiman 2011, pp. 600–601)

One could of course question the morality of the various decisions (on the part of insurance companies, hospitals, etc.) which have brought about a reduction in the quality of care. Nonetheless these are decisions which many views as socially acceptable in the United States, where the healthcare industry is profit-driven. For better or worse, it is an integral part of the healthcare community that members have "the ability to pressure other members of the community to accept more responsibility for providing healthcare" (ibid., 601).

If this sort of activity is acceptable when undertaken by certain members, such as insurance companies and hospitals, why should it nevertheless be unacceptable for other members to take similar action, such as nurses and physicians? Consider two courses of action: The first course of action is that insurance companies universally

[2]For similar positions, see Brecher (1985) and Grosskopf et al. (1985).

deny coverage for a certain medical procedure. The second is that physicians refuse to perform the same medical procedure. Since these actions would result in equivalent impacts on patient care, they ought to be given the same moral evaluation—they are either both permissible or both impermissible. As Neiman argues, this point holds more generally for the moral evaluation of strike action on the part of nurses (and physicians):

> If, from the viewpoint of the community's responsibility for healthcare, insurance companies are justified in making a choice, independent of the impact on patient care, to raise premiums, or hospitals are justified in making a choice, independent of the impact on patient care, to increase nurse-to-staff ratios, then nurses must also be justified in going on strike to protect their own interests or to advocate for the future of nursing, regardless of the impact on patient care… there seems to be no compelling reason why insurance companies, hospital administrators, and taxpayers should be justified in competing over their interests, but nurses should not. (ibid., 602)

Let me here mention two points that would have to be addressed before we could fully endorse Neiman's position. The first is the conditional nature of the defense of nurse (and physician) strikes. Neiman's argument is that *if* other members of the healthcare community may justifiably take action that negatively impacts patient care, then nurses (and physicians) may justifiably take such action as well. But of course it could be that such actions on the part of insurance companies, hospitals, etc., are not morally justifiable in the first place, in which case we could no longer say with confidence that nurses and physicians may justifiably take such actions.

The second point is that Neiman's position assumes that there is nothing special or unique about the role that nurses and physicians play in the healthcare system. But isn't it the case that nurses and physicians do have special duties when it comes to their interactions with patients—duties not borne by other members of the community? Neiman does of course argue that traditional consequentialist and deontological arguments for the special nature of the duties of nurses and physicians are deficient precisely because they fail to acknowledge the wider community's role in ensuring quality healthcare (ibid., 599). But acknowledging that other members bear some responsibility for ensuring quality care does not imply that nurses and physicians do not have more *stringent* duties to provide care; it merely proves that they are not the only members that have duties to provide care. These are each highly complex issues and I cannot hope to do justice to them here.[3]

Nevertheless I think Neiman is correct to identify the existence of a community responsibility to ensure quality care, part of which is to ensure adequate working conditions for healthcare employees. The system fails not simply when patient care is impeded but also when working conditions are poor—and the latter can of course impact the former. It is for this reason that nurse and physician strikes are their most justifiable when the key outcome of the strike is improved patient care, rather than

---

[3]For more discussion, especially concerning the purported special status of nurses and physicians, see MacDougall (2013).

(merely) improved nurse and physician standard of living. However, strikes should not compromise emergency care or other critical cases—nurses and physicians must make arrangements to ensure this kind of care in advance of a strike.[4]

Now even if the central assumption of my (and Neiman's) position is correct—that the provision of healthcare is a community obligation—it could still be that our moral evaluation of member actions differ from member to member. As Muñoz argues, there is a fundamental difference in how we hold nurses and physicians accountable and how we hold hospitals and insurance companies accountable:

> …people may get sick because of the actions or omissions of politicians and administrative bureaucrats. Politicians who, for instance, made bad decisions about the allocations of vaccines may harm many people. But the best way to hold politicians accountable for their actions and omissions is not by denying them freedom to strike, but by holding them electorally accountable… a physician's job is not to assign responsibility and act according to that judgment, but simply rather to treat her patients. (Muñoz 2014, p. 663)

The trouble here is that the election cycle may not turn quickly enough to address the injustices which might motivate a nurse or physician strike, and there is no guarantee that nurses and physicians could otherwise rally the public to their cause. Moreover, even politicians with the best of intentions still might find it difficult to coerce unelected members of the healthcare community—hospital administrators, insurance companies, etc. Perhaps over the long run patients, by making alternative insurance and hospital choices, could exert the kind of pressure needed to bring about improvements in hospitals and insurance companies. But many patients lack this kind of freedom concerning their healthcare choices.

It is an overstatement to suggest, as Muñoz does, that in deciding to strike, nurses and physicians would be assigning blame in an inappropriate way. Generally it is severely inadequate working conditions (to the point of compromising patient care) that motivate physician and nurse strikes—conditions for which the nurses and doctors are presumably *not* responsible. And since they are not responsible then responsibility for improving working conditions and patient care quite clearly falls on the other members of the healthcare community. Blame need not be assigned to any particular other member, except insofar as the goal of the strike is to address particular working conditions (in which case blame might fall on hospital administrators) or to address the availability of specific forms of treatment (in which case the blame might fall on private insurance companies). Physicians and nurses indeed have a duty to treat their patients, which, in extreme situations, might involve advocating for their patients by taking strike action. Thus, if other members of the healthcare community have placed nurses and physicians in a situation where they cannot effectively treat their patients, they may for this reason be justified in striking.

---

[4]These are conditions imposed by all authors who support (in-principle) nurse and physician strikes. See, e.g., Chadwick and Thompson (2000), Daniels (1978), Muyskens (1982), and Veatch and Bleich (1975).

To summarize, the position elaborated here endorses the permissibility of nurse and physician strikes when the following conditions have been meet:

1. That the provision of healthcare is the responsibility of the entire healthcare community and not exclusively the responsibility of nurses and physicians;
2. that the actions of the other members of the healthcare community are responsible for the injustices suffered by nurses, physicians, and patients;
3. that a major outcome of the strike is improved patient care; and
4. that arrangements are made to ensure emergency care during the strike.

I also believe that these conditions, suitably generalized, provide an intuitively appealing model for discussing the permissibility of other essential service strikes. To establish the permissibility of a strike for essential service *X* we would have to know the following:

5. Whether the provision of *X* is the responsibility of a wider community or whether it is the sole responsibility of the potential strikers;
6. whether the actions of the other members of this community (should they exist) are responsible for the injustices suffered by the workers or those benefiting from *X*;
7. whether a major outcome of the strike would be the improved availability and/or quality of *X*; and
8. whether steps will be taken to ensure continued access to *X* for those who absolutely need it (e.g., those who could not survive the anticipated disruption of *X* due to the strike).[5]

Note that (3) and (7) are not decisive factors. The claim here is that strikes are *most justifiable* when these conditions are met. I am willing to grant that an essential service strike could however be justified even if improvements to the service are not sought, but only if the existing working conditions are especially objectionable. This reflects the idea, identified above by Railton, that we cannot accept a system that requires some to suffer for the benefit of others. Just how much suffering is acceptable is of course a crucial question, but one that I will not address in any satisfying way here. However, I think we now have enough to go on to reexamine the issue of the CO_2 scrubber technician strike in Peary Crater City.

---

[5]I have intentionally omitted a further condition stipulating that the strike is only justifiable if it is successful. Strikers cannot plausibly predict the extent to which other members of the community will concede to their demands. Nonetheless if there are compelling reasons for believing that the strike will end in failure (because the demands could not possibly be met), then it would be much more difficult to justify the strike, as any harm caused during would not lead to improvements post-strike.

4.3 The Permissibility of Lunar Essential Service Strikes

To summarize the relevant details of the case described at the beginning of this chapter: The CO_2 scrubber technicians, who already consider themselves overworked, have just learned that their employer, LOST Inc., plans to implement layoffs and increases in worker responsibilities. But CO_2 scrubber technicians are not given sufficient time and resources to satisfy their already existing responsibilities and many of them have been docked pay for not finishing all of their daily assignments. LOST Inc.'s policies would, the CO_2 scrubber technicians believe, merely result in more docked pay and fewer responsibilities satisfied. As there is already a shortage of CO_2 scrubber technicians, layoffs are likely to threaten the quality of the air supply, possibly resulting in injury and death. In response the CO_2 scrubber technicians have gone on strike to increase their numbers and improve their working conditions.

Are the CO_2 scrubber technicians recklessly putting citizens' lives at risk, or are they making a desperate but necessary move to bring their working conditions up to reasonable standards? The conditions laid down at the end of the previous section should help us to answer this question.

Does the strike satisfy condition (5)? That is, does the service they are withholding, CO_2 scrubber maintenance, fall under a general community obligation? In the very least the provision of CO_2 scrubber maintenance is the responsibility of LOST Inc.—the technicians are only responsible for maintenance to the extent they can provide it under the conditions of their employment. But moreover, given the vital nature of air production, we can reasonably assume that wider social decisions play a role in how this industry is run. Let us envision some additional background information about the air industry. In particular, imagine that Peary Crater City law mandates the following "rough guidelines" for the air industry:

9. That, in order to err on the side of exceeding demand, the industry as a whole must produce air within a specified range of quantity and within a specified range of rates;
10. that, in order to foster efficiency via competition, a specified minimum number of providers must be active;
11. that, in order to prevent service disruptions, producers agree to meet individual air production quotas, as well as to establish excess production capacity for emergency situations;
12. that producers agree to and abide by specified air price ceilings; and
13. that producers agree to provide air containing industry standard mixes of gases.

These are certainly not the only regulatory principles we could envision as binding the lunar air industry, but it is safe to presume that some or all of these would form the basis of the "social contract" a lunar democratic society might make with the air industry. Assuming that these regulations are mutually agreed upon—that is, they are not dictated by the air industry—society is therefore at least partly responsible for the way the air industry is run. In particular, society is at least partly to blame, and thus is partly responsible for recompense if, under regulation, air production is only profitable through especially exploitative labor practices.

One can easily imagine the following context for the CO_2 scrubber technician strike action: New administration has been elected on promises of reducing the price of air. This has caused air producers to tighten their belts, resulting in LOST Inc.'s decision to cut corners as far as scrubber maintenance is concerned. Thus, wider social pressures are what started in motion a sequence of events that led to the unfortunate situation at LOST Inc. It would not be unreasonable to describe this situation as one in which society has demonstrated a preference for lower air prices over a reasonable standard of living for air production workers. But if there is a wider social obligation to produce the resources necessary for life then there must also be a concomitant obligation to make sure that no one is made to live an unnecessarily miserable life in the process. Society's preference for lower prices is in this case a violation of the obligation to make sure that no one is made to live an unnecessarily miserable life, and is therefore wrong. The CO_2 scrubber technician strike may be the only way of putting the pressure back on society. Therefore we can at least envision a possible way in which the CO_2 scrubber strike satisfies (5)—air production, like healthcare, is a community, and not an individual, responsibility.

It is clear that the strike also satisfies condition (6). The CO_2 scrubber technicians are not themselves responsible for their working conditions—those conditions are dictated directly by their employer, LOST Inc., and indirectly by wider social decisions regulating the air industry.

Whether the condition (7) holds is more difficult to determine. In the envisioned situation the strikers are not described as expressing any great concern for the quality and availability of clean air, rather, their concerns reside exclusively with their own working conditions and well-being. To this extent, then, we might be critical of the strike. However their poor working conditions appear to be symptoms of wider problems with the air production industry. In so far as their working conditions prevent them from completing all of their assignments, to continue or to exacerbate these conditions would mean a greater number of assignments left incomplete each day. Provided that enough of the CO_2 scrubbers are left unmaintained, the rising toxicity of the air will become a problem whether or not the technicians strike—all the strike will do, in the short term, is hasten the rise in toxicity, forcing the wider community to address the problem sooner. In other words, the existing working conditions of CO_2 scrubber technicians are a threat to the well-being of Peary Crater City. However, if the strikers' demands are met, the result would be an increase in the quality and availability of air. Thus a key outcome of the strike would be an improvement for the public over the pre-strike conditions.

That the strike would, if successful, benefit society as a whole is, I take it, the most salient point. It matters naught whether the technicians enter into the strike with narrow self-interest or with broader social concerns in mind. What matters is whether the working conditions raise broader social concerns, and in the described situation it is clear that they do. The narrow self-interest of the technicians may make it more difficult for them to win public support, which is of course relevant to the question of whether the strike is at all likely to succeed. Nevertheless the key moral question is not, what are the intentions of the strikers, but rather, would it be better, all things considered, if the strike succeeded?

I would also argue that a strike could still be permissible even if it failed to meet (7). Suppose that the CO_2 scrubber technicians' deplorable working conditions did not compromise their ability to adequately perform their jobs, and that meeting their demands would not improve the quality and availability of clean air. Still, the injustices faced by the CO_2 scrubber technicians warrant redress, and the overall societal cost of meeting their demands (for example, in slightly higher air prices) would be a small price to pay for vastly improving the lives of the technicians. Thus (7) could be relaxed to the more permissive:

7′. whether a major outcome of the strike would be *either* the improved availability and/or quality of *X or* the mitigation of strikers' genuinely concerning injustices at comparatively modest societal costs.

The envisioned working conditions of the CO_2 scrubber technicians are intended as a genuinely concerning injustice—but were they instead striking to receive one additional day of paid vacation per year then we should deem this as an "injustice" that fails to warrant a disruption in air maintenance work. This of course raises questions about which worker burdens count as genuinely concerning injustices and about which concessions count as comparatively modest societal costs.[6] It is difficult to say with much precision what counts here and what does not, and unfortunately I do not have space to sift through these issues in any great detail. However it may be that some existing theory of distributive justice, such as Rawls', is appropriate for this task (in which case our response to the workers' conditions and demands could be formed via the kind of idealized deliberation that Rawls envisioned).[7] The reader should be aware, then, that the coherence of the position defended here relies on there being *some* background conception of justice that could be used to identify which injustices are worth striking over.

The situation is not sufficiently specified to determine whether the last condition (8) holds—whether the strikers will make sure that emergency cases are covered. If, for instance, there is equipment in vital need of repair, such that, should this go unrepaired during the strike, lives will be lost or put at significant risks, then the strike is unjustified. However, if there are no such cases, and that the gravest consequences of the strike will be the delay in the routine maintenance and repair of CO_2 scrubbers, then the risks of the strike are, so long as it is not indefinitely protracted, rather minimal.[8]

[6]Thanks to Kelly Smith for bringing these issues to my attention.

[7]For more on the application of Rawlsian deliberation to space, see Schwartz (2015) and Milligan (2015).

[8]We might not be able to say the same concerning other positions in the air industry, for instance those tasked with monitoring or maintaining the air *supply* systems.

The strike action of the imagined CO_2 scrubbers appears to meet the conditions necessary for the strike to be morally permissible. It is therefore possible for there to be morally permissible essential service strikes, even in an industry as vital as air production. Of course, the imagined case was designed to sustain this judgment. Whether actual strikes will be justifiable will depend on a wide array of factors that we cannot at present control for: the size of the society and its air production industry; the specific means society employs for air production, distribution, and maintenance; etc. However, one detail that I take to be *irrelevant* is whether the industry is publicly or privately owned; a person's status as a *public* employee is not in itself of moral significance. Whether the air industry is publicly or privately owned, society as a whole remains a member of the air production community, and thus remains (partly) responsible for the well-being of air production workers. If the workers were suffering comparable injustices, and if those injustices posed comparable risks to the public, then we should give comparable moral evaluations for strike action.

However we should not ignore the concern that workers like the CO_2 scrubber technicians, if they are permitted to strike, might do so in impermissible circumstances. As Muñoz claims of essential service strikes,

> One the one hand, the cost of retaining essential service workers might in many cases become unaffordable. On the other hand, essential service workers could become incentive seekers who exploit the essential character of the service they provide. Restrictions to the freedom to strike are precisely justified to avoid this very problem. (Muñoz 2014, p. 668)

Here I do not disagree, in fact I think it is vital to the permissibility of air industry strikes that these concerns do not apply. However I do not think the "unaffordability" of the strikers' demands is necessarily a decisive impediment to the strike. So much here depends on how the society is organized and how the lives of the strikers compare to the lives of average citizens. The strikers demands could be unaffordable because the society genuinely cannot afford to provide *anyone* with a reasonable standard of living. But the strikers' demands could be unaffordable because society has decided to allocate its resources in an unjust way, and refuses to reallocate funding away from less morally worthy pursuits. Here the unaffordability is not *tout court* but qualified—it is not that the resources do not exist but rather that society has decided to provide the industry with less resources or support than the strikers find necessary. Which side to take here will turn on the overall justice of the existing system, perhaps using Pareto optimality as a criterion of justice or fairness. If the existing system does indeed bring about rectifiable injustices in the strikers' lives, then society has an obligation to alter its priorities, thereby shifting what it views as affordable. To reiterate Railton's comment, we cannot tolerate a system that makes the suffering of some a *requirement* of satisfying the needs of others. If we cannot do this, then we are not ready to thrive in space. Moreover, that essential service strikes are only permissible when they seek to rectify genuinely concerning injustices allows us to block (at least in principle) worries about incentive-seeking strikes. According to the position described here, non-justice economic concessions would not be an appropriate ethical basis for air industry strikes.

4.4 Young on Space Worker Strikes

Though to my knowledge the ethical discussion of essential service strikes in space *societies* is entirely novel,[9] nevertheless an ethical analysis of space worker strikes has been given previously by Young (1988). It should be useful to compare our positions to see if we occupy any common ground.

Young's discussion concerns solar energy employee strikes on O'Neill-type habitats. It is part of the scenario Young envisions that these habitats are constructed and maintained for the purpose of beaming solar energy back to Earth, and the habitats are in effect large "company towns" staffed by employees (who are perhaps joined by their families). Young assumes in addition that:

> Spaceworkers'...skill level...is likely to be much higher than the American average. Consequently they will be able to command a higher price for their labor than the American average. So the workers' ability to individually bargain for higher wages will be higher than normal. (ibid., 122)

In other words, Young is not contemplating undereducated blue collar workers but rather highly educated engineers and managerial staff. Moreover, given the high costs of constructing orbital habitats, Young assumes that worker expenses will be a very small percentage of the solar energy company's expenditures, and so

> Wages will be a relatively minor expense. Yet, to shut down the service, i.e., the "beaming" of electricity, will cost the company a great deal, while its maintenance and capital costs will continue to be high. Accordingly, the company stands to lose a lot by not giving into workers' demands, and not to lose very much by giving into them. (ibid., 123)

Thus, the spaceworkers that Young envisions have a considerable amount of power and control over their working conditions. This entails, Young suggests, that these workers have a considerable duty to refrain from taking advantage of their employer, which, as he describes, stands to lose much by not meeting worker demands.

I am skeptical on this point—even highly skilled technicians are often at the mercy of their employers, and there is no guarantee that the market will correct for less than ideal employment conditions. Things appear worse now than they did thirty years ago when Young's paper was written. Nevertheless I am willing for the sake of argument to grant Young's point. What is the upshot?

[9]Though of course the topic has been broached in science fiction, e.g., Robert Heinlein's *The Moon is a Harsh Mistress* and *Misfit*; Larry Niven and Jerry Pournelle's *Spirals*, as well as Allen Steele's Rude Astronauts series. Also worth mentioning is the *Babylon 5* episode "By Any Means Necessary" (Season 1, episode 12), in which the main plot concerns Commander Sinclair's response to a dock workers' strike. I thank Stephen Baxter for literature suggestions.

> A policy of strikes by workers (and, of course, concession by management) would, then, amount to an income redistribution policy which would take from the less fortunate of earth and give to the space elite. This seems wrong. Moreover, a strike would disrupt an essential service—the production of energy... To redistribute income from the (relatively) poor to the (relatively) wealthy, to interfere with essential services, and to jeopardize lives suggests a strong case against a policy of strikes. (ibid., 124)

Given the assumptions built into Young's scenario I am inclined to agree that a strike in such circumstances would be difficult to justify, however doubtful it is that I think such a situation might arise. The situation Young describes is one in which the strike will neither improve the availability and affordability of solar energy nor rectify a genuinely concerning injustice, but rather is the kind of incentive-seeking strike that Muñoz argues is impermissible in the essential services.

Nonetheless Young is willing to recognize situations under which it would be permissible for spaceworkers to strike:

> *Presumably, if working conditions were hazardous and management was jeopardizing the public interest in some way, the space workers could strike, in order to stop the company's interference with the rights of others. But otherwise, there seems to be a strong presumption against strikes by the spaceworkers.* (ibid., 125; Young's emphasis)

Young appears then to settle on a position on solar energy worker strikes that coheres with the framework outlined in Sects. 4.2 and 4.3.

It is however unclear what sort of general position Young is committed to. For instance, Young does not say whether the dual conditions in the solar energy worker case—hazardous working conditions and a jeopardized public interest—are individually or jointly necessary for the strike to be justified. It is possible to interpret Young as claiming that both conditions must be met in order the strike to be justified. If so, then it is possible that a generalized version of his position might only allow strikes in cases where genuinely concerning injustices affect *both* the workers *and* the wider public. It would then follow that my position is somewhat more tolerant of worker strikes than Young's, since I am willing to defend the permissibility of a strike when either of his conditions obtain.[10]

4.5 Conclusion

So is the envisioned strike actions of the LOST Inc. CO_2 scrubber technicians morally permissible? I believe the answer is yes, granting the additional assumptions supplied in Sect. 4.3. Thus at least some extraterrestrial essential service strikes are permissible, at least in some circumstances. We should not however hasten to the judgment that all such strikes are permissible. Rather, the upshot of

[10]Though it should be acknowledged that hazardous working conditions tend to be linked causally to service disruptions, so it might be that whenever the first condition holds the second holds (or is likely to hold) as well.

this chapter is that it outlines a framework that can facilitate deliberation concerning the permissibility of proposed strike (and other collective bargaining) action.

But as is always the case in applied ethical settings, the best normative principles are only as effective as those wielding them. The details of each individual case will always play a marked role in shaping such deliberations, and deliberators will need to have well-honed faculties of ethical judgment. It is undoubtedly the case that the resolution of many of the ethical questions about life in space posed here and elsewhere will have to wait until such time as we have adequate experience living in space. There is still however great value in contributing to and developing a stock of ethical concepts, principles, and theories that future space dwellers might wish to appropriate, and I hope I have made some small contribution in this regard.

A secondary lesson is that we ought to find a way to ensure that vital resources such as air are produced in the most autonomous and reliable of ways, so that no group of aggrieved individuals would have the power to place the well-being of an entire settlement at risk. This dovetails with much of what Charles Cockell has said about the so-called "problem of oxygen." Cockell's work aims primarily at mitigating the more institutional and governmental forms of tyranny (such as dictatorial or oligarchic control over vital resources), rather than my focus here, which has been on what might described as individual or "blue collar" tyranny. But one of Cockell's recommendations is nonetheless salient in the present setting:

> …multiplication of oxygen producing and distributing entities has a very prosaic and quite pragmatic advantage that has no relation to political philosophy or economics—it will provide redundancy in production during an emergency or failure of a major producer, preventing catastrophic reliance on a single entity. (Cockell 2013, p. 93)

Redundancy and multiplication of air production have additional pragmatic advantages: First, they are beneficial to society. If there is sufficient redundancy of production, maintenance, and distribution of air, then the strike actions of a single producer's employees will have less-than-devastating consequences for the society. Second, they are beneficial to workers. If workers are aware that strike action would not jeopardize the survival of the society, then they will feel more free to strike (and to engage in other collective bargaining activities) to rectify any injustices they face. Moreover, the strike actions of one producer's employees could in turn raise standards across the industry.

References

Braithwaite, S. (2000). Collective actions by physicians that do not endanger patients. *Cambridge Quarterly of Healthcare Ethics, 9*, 470–482.

Brecher, R. (1985). Striking responsibilities. *Journal of Medical Ethics, 11*, 66–69.

Chadwick, R., & Thompson, A. (2000). Professional ethics and labor disputes: medicine and nursing in the United Kingdom. *Cambridge Quarterly of Healthcare Ethics, 9*, 483–497.

Cockell, C. (2013). *Extra-Terrestrial Liberty: An enquiry into the nature and causes of tyrannical government beyond the earth*. Edinburgh: Shoving Leopard.

Daniels, N. (1978). On the picket line: Are doctor's strikes ethical? *Hastings Center Report, 8*, 24–49.
Dworkin, G. (1977). Strikes and the national health service: some legal and ethical issues. *Journal of Medical Ethics, 3*, 76–82.
Grosskopf, I., Buckman, G., & Garty, M. (1985). Ethical dilemmas of the doctor's strike in Israel. *Journal of Medical Ethics, 11*, 70–71.
MacDougall, D. R. (2013). Physicians' strikes and the competing bases of physicians' moral obligations. *Kennedy Institute of Ethics Journal, 23*, 249–274.
Milligan, T. (2015). Rawlsian deliberation about space settlement. In C. Cockell (ed.), *Human governance beyond earth: Implications for freedom* (pp. 9–22). Berlin: Springer.
Muñoz, C. (2014). Essential services, worker's freedom, and distributive justice. *Social Theory and Practice, 40*, 649–672.
Muyskens, J. (1982). Nurses' collective responsibility and the strike weapon. *The Journal of Medicine and Philosophy, 7*, 101–112.
Neiman, P. (2011). Nursing strikes: An ethical perspective on the US healthcare community. *Nursing Ethics, 18*, 596–605.
Railton, P. (1980). Health care personnel and the right to strike: A social perspective. In Marc Basson (Ed.), *Ethics, humanism, and medicine: Proceedings of three conferences sponsored by the committee on ethics, humanism, and medicine at the university of Michigan, 1978–1979* (pp. 291–302). New York: Alan R. Liss Inc.
Schwartz, J. (2015). Fairness as a moral grounding for space policy. In C. Cockell (ed) *The meaning of liberty beyond earth*, (pp. 69–89). Springer, Berlin.
Veatch, R., & Bleich, D. (1975). Interns and residents on strike. *Hastings Center Report, 5*, 7–12.
Weiss, M. (2000). The right to strike in essential services under United States labor law. In P. K. Villalobos (ed.), *Relaciones Laborales en el Siglo XXI* (pp. 95–196). Instituto de Investigaciones Juridicas de la UNAM.
Young, F. (1988). Labor relations in space: An essay in extraterrestrial business ethics. *The Monist, 71*, 114–129.

Chapter 5
Power and Discipline in Extra-Terrestrial Communities

Andrew Thomas

Abstract Lovecraft's school of "Interplanetary Fiction" allows novelists to consider freely extra-terrestrial human communities and their relation with Earth. Foucault's analysis of social order based on an understanding of power allows that this relationship may take many un-Earthly forms. If modern societies such as "Mars One" rely on personal identities within a community being affirmed through communication (such as being Nodes on the Internet), new structures of power and discipline may evolve. This paper seeks to apply a Foucaultian perspective on discipline in an extra-terrestrial community.

Keywords Space community · Space governance · China and moon · Mars community

5.1 Introduction

In this view of discipline in extra-terrestrial communities, a short debate about two contrasting approaches to foreign policy—the Constructivist and Realist approaches—offers a useful starting-point.

The view has been succinctly expressed by Wallerstein that (2004, pages x, xi):

> "…We have studied, in separate boxes to which we have given special names – politics, economics, the social structure, culture – without seeing that these boxes are constructs more of our imagination than of reality."

Wendt (1995) describes the Constructivist view of International Politics as comprising ontological and epistemological issues—by ontological he means that

A. Thomas (✉)
Department of Politics and Public Policy, Faculty of Business and Law, De Montfort University, The Gateway, Leicester LE1 9BH, UK
e-mail: Andrew.thomas@dmu.ac.uk

C.S. Cockell (ed.), *Dissent, Revolution and Liberty Beyond Earth*, Space and Society, DOI 10.1007/978-3-319-29349-3_5

"…social structures are collective phenomena that confront individuals as externally existing social facts."

and

"The epistemological issue is whether we can have objective knowledge of these structures."

Building on this, Slaughter (2011) succinctly defined the Constructivist approach to international relations in these terms:

"Constructivism is not a theory, but rather an ontology: A set of assumptions about the world and human motivation and agency…the variables of interest to scholars – e.g. military power, trade relations, international institutions, or domestic preferences – are not important because they are objective facts about the world, but rather because they have certain social meanings…constructed from a complex an specific mix of history, ideas, norms and beliefs which scholars must understand if they are to explain State behaviour"

She also defines "Realism" as:

"…for realists…the international system is defined by anarchy – the absence of a central authority…State power is the key-indeed the only – variable of interest, because only through power can states defend themselves and hope to survive." (2011).

This paper therefore paints on a broad canvas that is defined at least as much by our perceptions of it as by the actual impressions of the paint, as it were.

5.2 An Extra-Terrestrial Community: China and the Moon

Seeking an example of an extra-terrestrial community, in November 2014 the China Science and Technology Museum (中国科学技术馆) in Beijing offers on an upper floor an exhibit of the Moon.

In this empirical exercise, examining exhibits in the Museum, the three subject positions of the display space (Bennett 1995; Rose 2012) were identified: Patrons; guardians and curators; and visitors of the gallery; all interacting, and a "shooting script" (Suchar 1997) established to photograph the interaction and to repeatedly interact with the gallery in order to interrogate the transaction space.

There was no obvious guardian or curator in sight, which suggested some form of TV surveillance.

This artefact was not labelled, but was of the form of a large table at adult waist height upon which a model landscape and model structures had been placed. It was placed against the corner of the room; the walls of the room were painted blue with white "stars" and there was an image of Earthrise.

The display was analogous to a large model train set, except without any moving items.

Clearly a Patron of the gallery had established this model to demonstrate a view of what a Chinese settlement on the Moon would look like.

A portable camcorder was used to record views of this model. Specifically, labels in English and Chinese which identified facilities at points on the table were recorded. The moving image on one of the explanatory visual terminals was also re-recorded. Later the views were spliced together into a sequence and a start text and an end text were added.

This video clip is therefore a production, an interpretation, a construct of that model of a moon base.

One set of eyes looking at the constructed video could that of Hickman (2012) who considers "How Plausible is Chinese Annexation of Territory on the Moon?" and takes a realist view of it (p. 85):

> "Theories of realism or neo-realism, which explain foreign policy decisions as the result of interactions between decision-makers who are highly sensitive to the relative power of their states in an essentially anarchic international system, lend credence to the Chinese Moon scenario".

He considers (p. 86):

> "That China intends a landing on the Moon to be followed by the establishment of a permanent base for the purpose of enhancing its prestige is not a controversial contention. That it might also assert territorial sovereignty on the Moon has not been the subject of much speculation…

And (p. 85):

> This article examines three factors influencing China's decision of whether to accept the risk of asserting territorial sovereignty on the moon: (1) geopolitical motivation (2) strategic surprise and (3) international law. A plausible case for such a decision can be revealed from an examination of the three factors".

5.3 The Use of Science Fiction

This idea of annexation of extra-terrestrial territory can be explored in fiction.

H. P. Lovecraft's school of "Interplanetary Fiction" follows rules in which the plausibility of the story is carefully crafted. He writes (1975, p. 120):

> "Inconceivable events and conditions form a class apart from all other story elements, and cannot be made convincing by any mere process of casual narration. They have the handicap of incredulity to overcome; and this can be accomplished only through ha careful realism in every *other* phase of the story, plus a gradual atmospheric or emotional building-up of the utmost subtlety. The emphasis, too, must be kept right – hovering always *over the wonder of the central abnormality itself.*
>
> "Over and above everything else should tower the stark, outrageous monstrousness of the one chosen departure from Nature".

This is the necessary *Grundstein*, the foundation stone of science upon which the novel is built.

In *Of Martian Lines* (Thomas 2011), which was reviewed in the BIS' SF Journal Odyssey, the central stone of the plot is revealed at the end of the novel to be the annexation of Mars (pp. 195–6):

> "When the Russian Space Agency released the first pictures of the landing area and the plans for the colony that had been downloaded – with some effort as the dish had moved when the Command module had toppled – the Prime Minister addressed the Duma.
>
> Delegates were aware of the plans of the city that had been circulated, and knew of the Monument to Cosmonaut Valery Popovich, in the centre of a zone two versts wide, called "Earthly Fields", Zemlya Pole. They saw, too, that two long prospects had been marked out on the ground: Maxim Nikolayevich prospect, to the North, and Sergei Nilovitch Prospect, to the South, towards the mountain. The central administration block, which in time would be the site of the State lands Commission, was located in Zemlya Pole.
>
> "Our heroes of the Cosmonaut Corps," she said, "went to Mars to live, and to die.
>
> "Their deaths came too soon, and we mourn their passing.
>
> "But in their life was their intention to colonise Mars in the name of the Russian Democratic Federation.
>
> "They died with the wish that their new city, which they had started to lay out, would be part of our Federation.
>
> "Representatives of the State Duma! I ask you to accept their request.
>
> "I propose that the State of Mars, state capital Tsander City, shall be admitted at their request to the Russian Democratic Federation, and shall accept the laws and protection of our Federation".

So in this novel the entire planet of Mars would join the Federation. Why stop at just an occupied piece of land? It is as valid a construct as the annexation of the territory within a lunar base. The Realist view is only one of the possible views embraced as constructs.

Whatever ulterior purpose the colony or settlement may have, novels also provide an indication of life within the habitable structure.

The former President of the British Interplanetary Society, Arthur C. Clarke considers aspects of discipline and punishment almost, as it were, in passing, in his novel *Earthlight* (Clarke 1955, 1957), which is an adventure set on the Moon (p. 27):

> "Apart from the fact that its members were of distinctly superior intelligence, the Observatory was a microcosm of earth itself. With the exception of murder (and *that* was probably only a matter of time) almost everything that happened in terrestrial society was going on somewhere here. Sadler [a character] was seldom surprised by anything, and certainly not by this. It was merely to be expected that all six of the girls in computing, after some weeks in a largely male community, now had reputations that could only be described as fragile".

Clarke sees see social norms within the community and a dissipation of reputation within that community.

In another example he sees ulterior motives—perhaps he is in the Realist school—(p. 44):

> "On page four [of the newspaper] was a still more thought-provoking piece of news. A party of prospectors had been arrested on some remote asteroid in the vicinity of Jupiter. The charge, it seemed, was a violation of space-safety regulations. Sadler suspected that the charge was phoney – and that so were the prospectors. Central intelligence had probably lost some of its agents."

Here we see a system of extra-terrestrial law enforcement, an application of regulations and a suspected false motivation in the use of this enforcement system.

So in this fictional world, a world of constructs which includes, but is not limited to, that realist definition, we can see the importance of motivation, of trust, and of community norms, including reputation.

These are factors which are constructed around social interaction, and their application to discipline has been considered by Foucault.

5.4 A Perspective Owed to Michel Foucault

The political philosopher Foucault (trans. 1977) reviewed the philosophy of discipline and punishment and in particular the transition, which took place in Europe around the eighteenth century, from physical torment of the transgressor to his or her imprisonment and rehabilitation.

Foucault brings to the debate a distinction between two generations of Discipline. If these are considered to be the first two generations, the third generation is proposed to take place in extra-terrestrial communities.

Graphic images and actual devices are exhibited nowadays in museums, for example the National Museum of Denmark in Copenhagen, where a display describes a transgression as a breach of the Laws of the Old Testament and advises that "God was to be appeased through severe sentences aimed at retribution and disgrace".

Foucault (trans. 1977) writes (p. 35):

> "The tortured body is first inscribed in the legal ceremonial that must produce, open for all to see, the truth of the crime."

He describes his approach to the issue in these terms (p. 23):

> "This book is intended as a correlative history of the modern soul and of a new power to judge; a genealogy of the present scientifico-legal complex from which the power to punish derives its bases, justifications and rules, from which it extends its effects and by which it masks its exorbitant singularity".

Clarifying this, Schwan and Schapiro find (p. 25):

> "In other words, rather than assuming that we have always had a sense of our interior self, our soul, he will argue that this sense of personal identity has been socially conditioned and historically constructed".

This argument, of the social construction of punishment, and indeed of my perception of space exploration, stands in contrast to that of the realist school.

Foucault describes (page 14) the ending of physical torture:

> "At the beginning of the nineteenth century, then, the great spectacle of physical punishment disappeared; the tortured body was avoided; the theatrical representation of pain was excluded from punishment. The age of sobriety in punishment had begun.

And on page 16

"The reduction in penal severity in the last 200 years is a phenomenon with which legal historians are well acquainted. But, for a long time, it has been regarded in an overall way as a quantitative phenomenon: less cruelty, less pain, more kindness, more respect, more "humanity". In fact, these changes are accompanied by a displacement in the very object of the punitive operation".

Schwan and Shapiro (2011) explain this as:

"The important shift in the last two hundred years, then, is not the superficial reduction of penal severity and increased desire to respect the human rights of prisoners, but that punishment's *objective*, its object or target of operation has changed. Punishment no longer wants to handle the outside body but instead wants to get *inside* the prisoner and impact the prisoner's heart, mind, will and inclinations; in short, the criminal's *soul*.

The shift to a third generation of discipline and punishment in extra-terrestrial communities will be based on identifying the objective of punishment in that community and its means of changing the behaviour of the transgressor.

5.5 Foucault's Nodes and Networks

An extra-terrestrial community is considered as a community consisting of Nodes which acting together create a Network.

Foucault's idea of "Nodes" (trans. 1977, p. 138) is:

...this new political anatomy...is...a multiplicity of often minor processes, of different origin and scattered location, which overlap, repeat, or imitate one another, support one another, distinguish themselves from one another according to their domain of application, converge and gradually produce the blueprint of a general method".

Interpreting this, Schwan and Shapiro, explain (p. 102):

"...we are held in place by a capillary network of multiple small nodes, each of which contributes to our subordination, but which can also be compensated for if one fails or is dismantled"

There are many definitions of the word "NETWORK"; for example the definition in computer terms by OSI distinguishes several layers between the physical connection of cables and the application which sits at the very top for humans to use.

Social Network Analysis is a technique to show by a graphic the interactions of individuals with each other; there is a developing science on the topic (see for example Easley and Kleinberg 2010).

Foucault (trans 1977) gives a definition at pp. 176–7:

...although surveillance rests on individuals, its functioning is that of a network of relations from top to bottom, but also to a certain extent from bottom to top and laterally; this network 'holds' the whole together and traverses it in its entirety with effects of power that derive from one another: supervisors, perpetually supervised".

And on p. 177

"Discipline makes possible the operation of a relational power that sustains itself by its own mechanism and which, for the spectacle of public events, substitutes the uninterrupted play of calculated gazes. Thanks to the techniques of surveillance, the 'physics' of power, the hold of the body, operate according to the laws of optics and mechanics, according to a whole play of spaces, lines, screens, beams, degrees and without recourse, in principle at least, to excess, force or violence".

In "Nullitopia" (Thomas 2014) unknown fictional characters wear communicative headphones during the day to connect with social networks as Nodes, and are seen particularly lining the sides of the carriage of the underground train in Hong Kong as if they are artificial nodes forming together a neural network of an artificial snake or worm.

Foucault, famously, wrote of Bentham's Panopticon, a prison of surveillance Foucault (trans 1977, p. 201):

...the major effect of the Panopticon: to induce in the inmate a state of conscious and permanent visibility that assumes the automatic functioning of power".

He describes its organisation thus (trans. 1977, p. 202):

...in the peripheric ring, one is totally seen, without ever seeing; in the central tower, one sees everything, without ever being seen....It is an important mechanism, for it automatizes and disindividualises power".

For the third generation of discipline in an extra-terrestrial community is proposed a visual surveillance in which one consents to surveillance by participation in the surveillance network, which is in the guise of a self-policing social network.

5.6 An Example of the Third Generation Network: Mars One

The Mars One project provides a test-bed for theories of all kinds, including the function of discipline. Ralph (2013) reports on the thoughts of the founder, Bas Lansdorp (Ralph 2013):

"The international scope of the project raises interesting questions about the legal system that will be enforced on the Mars colony. Since most of Mars One's technology will be American-made (the United States is at the forefront of the world's missions to Mars), Lansdorp reasons, the astronauts could be said to be living on "American soil," and U.S. law should govern.

"But in practice, Lansdorp adds, governance will look more like it does on the International Space Station and at research stations in the Arctic, where a mission commander decides what flies and what doesn't. In the event of a disagreement with the mission commander, "ground control" will moderate, he notes.

"When there are just four people on the planet — all of whom have been training together for years — that system might work. But later on, as more people colonize the planet, will it hold? "We are on earth and they are [on] Mars," says Lansdorp. "And there's millions of kilometres between us. So, at some point, the people on Mars are going to say,

'We don't care what you say; we're going to do it our way.' When this will happen we don't know. It could happen when there are just 12 people; it could happen when there are 20 people; it could happen when there are 50 people. But at some point, they're going to declare their independence." And, Lansdorp makes sure to point out, "I think that will be awesome."

This reported discussion shows here not only the implications for foreign policy on Earth noted earlier, but the reference to the International Space Station, and the eventual aim of the Mars community to do things "Our way".

There has been some consideration of the form of computer network that might be built on Mars. The "Marspedia" project writes:

"The first permanent settlement on Mars will be of a size that does not necessarily require a distributed computer network. Nevertheless, there is the need for a mutual communication between people on Mars and people on Earth. The Martian settlement can, therefore, be connected to the terrestrial Internet, which in fact makes it an **Interplanetary Internet**. *(emphasis original)*

...

In the first instance (when the first colony is being constructed) it may not be possible to have a functioning "Internet" as we currently understand it. Colonists should, however, have email contact for personal and science-based contact with Earth.

On Mars One:

"Our astronauts will also find time to relax. They can do most of the indoor activities that people can do on Earth: read, play games, write, paint, work out in the gym, watch TV, use the Internet, contact friends at home and so on.

The settlement will be broadcasting images of daily life back to Earth so that everyone can see what the astronauts are up to"

This form of settlement is planned to a communal space:

"Construction involves working on the settlement"

"Our goal is to enable them to construct a space 10 meters wide by 60 meters long. This will be a spacious environment in which to live, where they can also grow trees. Such a large living volume will make Mars a much nicer place to live".

But communal structures built in situ may resemble the sort of ad hoc communal civil defence shown in the former Soviet Union's instructions, where a community digs a trench and bends saplings into rings to lower into it, forming a frame, which is then covered in soil or regolith.

5.7 Discipline Within the Communal Structure

Foucault (trans 1977, p. 92) is acutely aware of the transgression within the community:

...the injury that a crime inflicts upon the social body is the disorder that it introduces into it: the scandal that it gives rise to, the example that it gives, the incitement to repeat it if it is not published, the possibility of it becoming widespread that it bears within it. In order to be useful, punishment must have as its objective the consequences of the crime, that is to say, the series of disorders that it is capable of initiating…

An example of a community transgression, based on work with Astronauts on the International Space Station (Stuster 2014; Nimon 2011), can be suggested.

Stuster (2014) describes his "Journals Flight Experiment":

"Ten astronauts participated in the original study and wrote most about work, out-side communications, adjustment to the conditions, and interactions with crew mates. The study found evidence of a decline in morale during the third quarters of the expeditions and identified factors that contribute to sustained adjustment and optimal performance".

Nimon (2011) elaborates the study:

Based on crew writings, the investigation identified 24 major categories i with behavioural implications. Of these, there were 10 in particular that accounted for 88 % of the text in the journal entries. These specific focus areas included: work, outside communications, adjustment, group interaction, recreation/leisure, equipment, events, organization/management, sleep and food."

What were the difficulties experienced by astronauts? The German magazine *Der Spiegel* reported on them (Höflinger 2014)[1]

"Before flights to the International Space Station (ISS) asks Stuster NASA astronauts to describe their lives on board: how they feel ("decidedly morose") what they annoying (Houston), after which they yearn (tortillas).

They praise the camaraderie: "the best pal to spend six months in a tin", and sometimes even the food: "You send ice cream."

Clearly food and the monotony of it plays a large role in astronauts' lives. So as a concluding exercise the story in Box 1 is imagined about a transgression in a community on Mars:

Box 1: A social transgression on Mars

Jack is fed up with eating food which is a paste from a tube. He wakes up in the night and craves some cheese that he can chew. Being an enterprising sort of man, he goes to the laboratory area and switches on the 3-D printer. He designs a slab of cheese, and inserts the paste into the printer; it extrudes warm cheese which smells cheesy and sets on the tray.

The following morning the community information network, Radio Mars, contains a special bulletin: "Please do not print cheese with the three-dee printer". This gives rise to a lot of speculation within the community. Enraged, Jack creates a communal broadcast he called "Radio Rotten" in which he advises: "Do not feed the Mice".

[1]"Vor Flügen zur Internationalen Raumstation (ISS) bittet Stuster Nasa-Astronauten, ihr Leben an Bord zu beschreiben:… wonach sie sich sehnen (Tortillas)".

And

"Sie loben die Kameradschaft: "Der beste Kumpel, um sechs Monate in einer Blechbüchse zu verbringen", und manchmal sogar das Essen: "Sie schicken Eiscreme"."

The response from Radio Mars is the further comment: "There are no Mice on Mars". But Jack has gone too far and is admonished for his behaviour.

As a result, on the communal screen showing the location of all members of this small community, a screen which is copied to Earth, all the avatars are in the same colour.

Except one: Jack's avatar is now in a different colour, identifying his location within the community but transgression from its norms

Here, it is the public exposure and isolation of "Jack", and the stress, embarrassment and shame of being the 'different one', a transgressor—a "dodgy", untrustworthy person whom others should be wary of—that is the punishment; more so than any loss of privileges arising from the transgression.

5.8 Conclusion

In this paper, the importance of social constructs has been addressed together with the explanation of discipline by Michel Foucault, to address the issue of transgression in extra-terrestrial communities. The communal social network, which members consent to joining, functions to impose behavioural norms on members by its capacity to modify in public the terms of its inclusion, as the nodes on the network turn in on themselves. After the torture of the human body, and, later, its imprisonment, this is considered to be a third-generation system of punishment. The prisoners guard the guards.

References

Bennett, T. (1995). *As summarised in Rose, G. 2012. Visual methodologies* (p. 239). London: Sage.

Clarke, A. C. (1955, 1957). *Earthlight*. Pan Books Edition.

Easley, D., & Kleinberg, J. (2010). *Networks, crowds and markets: Reasoning about a highly connected world*. Cambridge: Cambridge University Press.

Foucault, M. (1977). *Discipline and punish—The birth of the prison* (A. Sheridan, Trans.). London: Penguin.

Hickman, J. (2012). How plausible is Chinese annexation of territory on the moon? *Astropolitics, 10*, 84–92.

Höflinger, L. (2014). Ich Will Heim. *Der Spiegel, 24*(2014), p115.

Lovecraft, H. P. (1975). Notes on writing interplanetary fiction. In *Supernatural horror in literature and other Literary Essays*. Rockville: Wildside Press (pp. 119–127).

Mars one (undated). http://www.mars-one.com/faq/technology/how-does-the-mars-base-communicate-with-earth; http://www.mars-one.com/faq/mission-to-mars/what-will-the-astronauts-do-on-mars. Accessed May 26, 2015.

Marspedia. Undated. http://www.marspedia.org/index.php?title=Internet. Accessed May 26, 2015.
Nimon, J. (2011). Journaling the journey into space. International Space Station Program Science Office, NASA's Johnson Space Center 12.19.11 http://www.nasa.gov/mission_pages/station/research/news/Stuster_Journals.html. Accessed May 26, 2015.
Ralph, E. F. (2013). http://foreignpolicy.com/2013/03/28/interested-in-a-televised-one-way-super-competitive-mission-to-mars. Downloaded May 26, 2015.
Schwan, A., & Shapiro, S. (2011). *How to read Foucault's discipline and punish*. Pluto.
Slaughter, A. M. (2011). International relations, principal theories. In R. Wolfram (Ed.), Max Planck Encyclopaedia of Public International Law.
Stuster, J. (2014). *Journals flight experiment*. NASA Human Research Program Investigators' Workshop (2014). Abstract: http://www.hou.usra.edu/meetings/hrp2014/pdf/3002.pdf. Accessed May 26, 2015.
Suchar, C. S. (1997). Grounding visual sociology research in shooting scripts. *Qualitative Sociology*, *200*(91), 33–55.
Thomas, A. (2011). *Of Martian Lines*. Amazon: Createspace.
Thomas, A. (2014). *Nullitopia*. Amazon: Lulu.
Wallerstein, I. (2004). *World systems analysis: an introduction*. Duke UP.
Wendt, A. (1995). Constructing international politics. *International Security*, *20*(1), 71–81.

Chapter 6
Space for Dissent: Disobedience on Artificial Habitats and Planetary Settlements

Mukesh Chiman Bhatt

Abstract Several historical instances of supposed rebellion or discontent in the various space programmes, on Skylab, Mir, Salyut/Almaz and lastly the ISS are presented and examined with respect to and similarly for proposed intra- and extra-solar settlements from a legal perspective, using current international agreements and the various ISS codes of conduct and disciplinary policies. Some analysis will require the use of international codes and examples from fiction.

Keywords Dissent · Rebellion · Mutiny · Outer space · Space stations · Intergovernmental agreement · Astronaut code of conduct · Astronaut behaviour

6.1 Introduction

> Space is huge enough, so that somewhere in its vastness there will always be a place for rebels and outlaws. Near to the sun, space will belong to big governments and computerized industries. Outside, the open frontier will beckon as it has beckoned before, to persecuted minorities escaping from oppression, to religious fanatics escaping from their neighbors [*sic*], to recalcitrant teen-agers escaping from their parents, to lovers of solitude escaping from crowds. Perhaps most important of all for man's future, there will be groups of people setting out to find a place where they can be free from prying eyes…
>
> FREEMAN DYSON, *From Eros to Gaia* (quoted in Steele 2002, p. 8)

Fight or flight: the two best known responses of social interaction between animals. Where two or more free thinking individuals meet, there is bound to be dissent and discord, a difference of opinion that may lead to change. This change may be a change of authority, a change of hierarchical structure and consequent change of status, or a change in the differential power of those involved. Most importantly, it may lead to a parting of ways, with subsequent colonisation of far off lands, with relations both commercial and familial being established later. Given the

M.C. Bhatt (✉)
School of Law, Birkbeck College, University of London, London, UK
e-mail: m.bhatt@physics.org

C.S. Cockell (ed.), *Dissent, Revolution and Liberty Beyond Earth*, Space and Society, DOI 10.1007/978-3-319-29349-3_6 constant presence of disputes, it is surprising that there is no known law of rebellion or revolution. However, the fight or flight paradigm used in animal behaviour may be used to present a possible basis on which to presume a universal norm of human action. This can be justified in later human and moral terms by appeal to higher authority, or to some universal law of nature.

There have been several instances of supposed rebellion and discontent on the various space stations to date: Skylab, Mir, Salyut/Almaz and the ISS. These will be presented, along with one example of unacceptable activity in a terrestrial field analogue, and the response of an astronaut, Chris Hadfield, in answer to a question about mutiny. The relevant sections and articles that govern the behaviour of persons aboard previous and current space habitats will be discussed, and some analysis attempted, with appeal to relevant analogues and parallels where possible.

6.1.1 Historical Background

The earliest records of the right to rebellion appear to be in either the Vedic (Hindu) cultures based in and around present-day Northern India and the Indus Valley, or in Northern China. Chinese rebels would claim the Mandate of Heaven, a doctrine which had been developed during the Zhou dynasty, to have been removed from an unjust or despotic ruler, thus approving revolution, in order to settle the Mandate of Heaven on another, supposedly more just ruler. It appears that, rightly in their view, Mencius (372–289 BC), a Confucian philosopher had his writings suppressed by later dynasties for declaring this right of the people to revolt against those rules that did not provide for the people's needs. Similarly, the Vedic attitude declares tyranny and unjust rule as contrary to *ṛta* and *dharma*, these terms denoting the structure and the order of the universe. This latter should be distinguished, strange as it may seem to Western political philosophers, from the exercise of authority by a sovereign, which went by the name of *ḍanḍanitī* (literally "conduct by the stick"). In both cases it should be noted, despite the translated terms, that there is no "god" or "gods" that have laid down this rule. The Vedic version is best known in the West from translations of the Bhagavad Gita, which is a set of philosophical discursions set in the midst of a battlefield, the primary purpose being to justify the war against unrighteousness about to be waged. Krishna, in Chap. 4, declares that unrighteousness must be opposed, based on which Gandhi and Tilak opposed the British in India, ultimately leading to independence for India. Paradoxically, despite their common basis for action, these two revolutionaries took opposing paths, the one non-violent action, the other violent revolution. Similarly, in later centuries, the French Huguenots and other, English Protestants also reserved the rebellion as a moral right within their interpretation of Christianity.

Whereas much rebellion was based on need, tyranny and disagreement, some was based on an inability to tolerate divers and adverse conditions, whether war, famine or other disaster. Invariably, however, the response of the authorities would have been to suppress dissent. It was rare in the early years of human civilizations and cultures to allow dissent that justified opposition as the Chinese and the Indians did. The early Middle Eastern, and the later European response was invariable: oppress, suppress, and punish. It is only later that conviction of belief and conscience became the basis for opposition to authority. This right of rebellion was then enshrined in a number of constitutions in the 18th and later centuries, the most famous of these being that of the French Revolution, and of the American revolution. Further examples with an American bias may be found in the Wikipedia article on the right to revolution (accessed 2015). A very interesting period with significance for the independence of artificial habitats and planetary settlements in future is the period of decolonisation in Africa from 1957 onwards, with especial reference to unilateral declarations of independence and the rebellions that preceded them. With the rise of the modern world and European dominance came structural and institutional mechanisms intended to restrain or suppress dissent, but which also allowed the expression of dissent under carefully controlled conditions, necessitating the collection of statistics and censuses, suitable for coercion, monitoring and the surveillance of populations.

Another interesting response of a segment of the population was to emigrate and establish a settlement, or to join an existing one elsewhere. This was standard practice in Vedic culture where fighting or rebellion was not a viable option, and may be considered important for settlements in space, subject to the restrictions placed upon these settlements by a hostile environment. The most famous example, having entered the mythology of the founding of America, is the emigration of the Pilgrim Fathers on the *Mayflower* to the new lands, supposedly fleeing religious persecution.

In international law, intergovernmental agreements, conventions, protocols, codes of conduct and standards all perform a similar function legitimizing rebellion, providing a regulatory and statutory framework for legitimate dissent and opposition often institutionalised in reality, through, for example, the Outer Space Treaties, the International Court of Justice, a proposed World Space Authority or the numerous non-governmental organisations such as the International Astronautical Federation, or the Deep Sea-bed Authority.

6.1.2 The Hostility of Space

There are two aspects peculiar to the environment of space that need to be considered at all times. Space is a hostile environment, and currently, all current space-going ventures are undertaken under a joint military-civilian command structure. That space is inimical to humans is well known, and mirrors the difficulty of living in difficult environments such as the Antarctic or under the sea. There can be little or no compromise—given the wrong set of decisions and conditions, the inhabitants, failing rescue, will very likely end up dead. Early settlers from Europe in the Americas had the benefit of living in a reasonably hospitable situation, wild animals and diseases notwithstanding; help was available in the form of the native Americans, and fresh air, water and food was available more readily than it would be in space, with the latter's added radiation and other hazards. In space, the essential resources of food, water and air need to brought in from outside. In order to avoid catastrophe, strict monitoring would be necessary, if not actual and continuous surveillance. This of course is likely to restrict the ability of the inhabitants to rebel in any way that might harm themselves or their fellow inhabitants, with the necessity of brute survival needs overriding all other concerns. Although exaggerated in space, similar situations do arise on Earth, namely in self-contained habitats established in hostile environments, in Antarctica, or in the deep sea. Parallels may also be drawn with oilrigs and aircraft carriers, which have to operate independently for long periods to time, or with the castles of Cathars and others under siege, placed upon high mountains and hills, all of which have limited supplies of food and fresh water.

6.1.3 Hierarchies in Space

Rebellion may be considered in any of these self-contained habitats. In general, there are very specific command structures and hierarchies in place in any of these situations. Current space ventures are both military and civilian in nature. Military command on aircraft carriers or in military castles under siege presents a simple hierarchy, in which mutiny can occur, primarily an opposition and refusal to follow the orders of a superior authority. It is therefore necessary to draw a distinction between these various types of command structure, and therefore the institutions that regulate the behaviour within these habitats. Mutiny, piracy, hijacking, rebellion and similar terms are however used interchangeably in this essay, although they have very specific technical meanings in given situations.

In those cases considered here, violence may occur and be acceptable, *provided it does not harm the population at large.* However, minimising harm requires prudence in the use of violence, non-violent methods of opposition being preferable.

6.2 Types of Dissent

There are several levels of dissent, those that exist between individuals and their difference of opinions, those that exist between an individual and a group to which that individual nominally belongs, those of the group as contrasted to the wider society and authority structures within which it is situated and finally that between the wider society within a global context, specifically that between nation-states. Although individual concerns are of necessity the basis for much dissent, in general these concerns are raised as part of a representative group within a larger context. Examination of the examples discussed within this descriptive essay remain largely within the context of the behaviour of individuals as representative of astronauts working within the context of the international space station and the civilian command structures in place for its governance. Where possible, this will be generalised and commented upon within the wider background that affects human interaction as various levels, both within the context of space, and within human society and its institutionalised disciplinary and regulatory structures. Consequently, there will be considerable overlap in the various types of dissent as examined. However, as stated, the primary focus of this work will be on the interaction between individuals (or groups of individuals) and the hierarchical structures within which they take issue with several concerns that affect these individuals directly. Interested readers are referred to Goh (2007) for a fuller discussion on settling disputes within the context of international space law at varying and different levels, including that between nation-states, commercial enterprises and international organisations. Disputes between nation-states regarding matters in space are primarily resolved by appeal to international law, the outer space treaties, and in the matter of liability, to the courts under the national jurisdiction of the nation-states involved at various stages of the space-going enterprise, with ultimate arbitration being vested in the international instruments of justice, the International Court of Justice. However, this is beyond the scope of this project.

6.2.1 Breaking the Law

A full discussion of motives for dissent or disagreements is beyond the scope of this essay. However, primary amongst these is a perception of mistreatment or experience of perceived injustice. Reasons for dissension may of course be the lack of help or support during natural disasters, war, famine and flood, the practice of tyrannical punishment or authoritarian coercion by elected or unelected representatives whilst in office. Often, the disagreement is of an ideological nature, which includes questions of ethics, religion and morality, with the nature and exercise of power, inequalities of distribution and lack of opportunity providing further fuel for opposition to the order of things. In all, the current paradigm in the Western (and therefore "modern") world is provided by the context of rights and responsibilities. The United Nations Universal Declaration of Human Rights (UN-UDHR) in its Preamble states that '…disregard and contempt for human rights have resulted in barbarous acts which *have outraged the conscience of mankind…*' with Article 1 continuing, 'All human beings are born free and equal in dignity and rights.' With regard of the assumed basis for these rights and responsibilities, there is a vast literature, but see Donovan and Anderson (2003) for an American perspective from an anthropological and legal point of view.

Dissent takes many forms, and evokes several different types of responses. At heart, as discussed by Brownlee (2012) are matters of conviction or conscience, or both. Dissent then takes the form of an unwillingness to follow rules and regulations or the law, to accept a difference of opinion, or some other reason. However, this can be contained and managed, paradoxically, by and within the very same rules and regulations allowing dissent and rebellion. Is this then rebellion against authority, or is acting in accordance with the law approved by authority not an act of dissent?

The effort to control dissent also evokes a set of responses: primarily violent and non-violent acts, but also migration or revolution, all necessitating acts of civil disobedience. Brownlee (2012), on whom part of the following discussion is based, sees in the standard liberal perspective the presence of conviction as personal, in contrast to civil disobedience as necessarily communicative, strategic, risky, and participatory, but undemocratic, and both being conscientious and conscience driven. Civil disobedience can however have a moral right, and also have a legal excuse for a breach of the law, with a rejection of the idea that such breaches of the law must be punishable, particularly in relation to laws that we oppose, even when these originate from valid normative sources. Of importance to the present discussion is Brownlee's argument (2012, Chap. 6) of the notion of necessity providing a legal justification for civil disobedience, "animated by a legitimate concern for people's non-contingent basic needs and rights … not only brute survival needs, but also the equally fundamental humanistic needs of basic political recognition, social inclusion, and respect" but which is not available to "bigots, xenophobes, and misguided do-gooders" (Brownlee 2012, ibid.). Whereas Rawls' view of civil disobedience as seen by Brownlee is that of a public, non-violent, conscientious and political breach of the law communicated in advance, and a willingness to accept punishment, this acceptance of sanction exhibits a loyalty to the law and the concept of justice. Given that violent acts of resistance draw attention to the violence rather than the cause, non-violent acts are preferable, but can also cause more collateral harm to others than violent ones. Prudence dictates non-violence, reducing the risk to life, of harm or of antipathy or harsh countermeasures. As discussed by Brownlee at this point, Raz shows that a breach of the law is committed in any act of civil disobedience provided there is a moral prohibition against that law, and that that is the very reason the law is breached. Brownlee cites the case of Gandhi as a civil, non-violent protestor, a revolutionary who willingly accepted jail many times. There may therefore be a *moral right* which includes the possibility of deliberately breaking the law. This is especially important in view of the inimical and hostile nature of the Outer Space environment, which poses significant risk to the average human. Consequently, although civil disobedience is supportive of non-contingent basic needs and rights, it is also a deliberate breach of law that can and often does cause damage or harm.

6.3 Dissent Aboard Space Stations

There is a little-known roster of astronauts who have rebelled when faced with heightened risks or tensions in exploring space. As is well known, there has been a series of manned space stations placed in orbit around Earth, primarily by the first two space-going nations, the USA and the then USSR, later Russia. Good histories of each of these are available elsewhere and the interested reader is referred to these, available readily by utilising search engines on the Internet. Although the Apollo series of missions are not technically a space station, these are some of the earliest examples of the short, transient habitation of space. The series of nominally civilian space stations starts formally with Skylab, then the Salyut habitats, which are simultaneous and synonymous with the USSR's Almaz military missions, followed by Mir, and finally the International Space Station, which is still operational at the time of writing. The relevant incidents that have been described as mutinies in each of these installations will be examined in turn, with such detail as is required to bring out the salient details of each. It should be noted that although the term mutiny is used, the incidents appear trivial when compared, say, to the mutiny aboard the Bounty, or similar events. Further, the command structures in each case are mixed, supposedly primarily civilian, although the astronauts involved are often military personnel under the command of a military hierarchy.

The data are taken from newspaper reports, and from extant documents available from various repositories. Unfortunately, the actual transcripts of incidents were not accessible to the author at the time of writing because of language and other restrictions. This particularly applies to those materials concerning Soviet-Russian expeditions, and to codes of conduct for Chinese Taikonauts and Soviet and Russian cosmonauts. I have been unable to date to establish whether similar documentation exists for the other space-faring nations, namely India, and the signatories that belong to the European Space Agency. However, these are also partner to the various treaties and agreements governing the International Space Station and to relevant UN instruments, and can be considered to adhere to the conditions specified in agreements concluded bilaterally or otherwise with NASA and the United States. However, this will result, as is often the case, in a strong bias towards the values espoused by the United States of America, although there is no intention to ignore the value systems of other cultures and countries.

6.3.1 Salyut

Salyut has had no incidents or events that have reported in the media or elsewhere as instances of mutiny or disobedience. If they exist I have been unable to find or access this information. As such, it is the easiest to discuss. It is possible that since several Salyut missions were concurrent with the broadly military and therefore presumably secret missions, the details of these remain classified. Any reader who has relevant information is invited to pass this information on the present author at the above email address, and these will be gratefully received and acknowledged.

6.3.2 Apollo 7

One of the first instances of defiance occurred with the Apollo 7 crew, in October 1968. An earlier Apollo capsule burned on the ground killing 3 astronauts; the goal of the 11-day Apollo 7 flight was to test the command and service modules preparatory to the Moon landing. The crew was therefore tense, feeling the burden of the future of the space programme. Worse, all three crew members got colds whilst in orbit, making Walter Schirra, the commander, especially testy. He subsequently refused to obey any orders from ground control (Broad 1997). Schirra, a pilot and Mercury 7 astronaut, snapped when a ground controller complained about his attitude; he responded with a "go to hell" (Fuentes 2015). Despite this, the mission was an unqualified success, and contributed to the continuation of the Apollo programme. No action is reported to have been taken against him for this minor breach.

6.3.3 Skylab 4

As the first example of a mutiny in space, the events on Skylab 4 in December 1973 have been duly reported several times over and are the ones with the fullest record. The following account is summarised from Broad (1997), Teitel (2013), Vitello (2014), Fuentes (2015), Kay (2014) and especially Freeman (2000).

Skylab was a series of missions to establish a space station, following the successful lunar landings. The first was the launch mission, and the crew of the second mission spent 28 days in space for repairs, the installation of a sunshade, and conducting space astronomy, Earth resources and medical study observations. The Skylab 3 crew spent 60 days on the station, an incredibly enthusiastic and efficient team who continued maintenance and experimentation. Skylab 4's crew endured a record 84 days in space from November 1973 to February 1974, following a strict schedule, which included 16-hour workdays. However as time went on, the crew fell further and further behind schedule.

Freeman (2000, pp. 7–9) describes the emotional and psychological effects of the isolation of spaceflight on Skylab 4. The crew were first-time rookies in space; when they endured space motion sickness they disregarded NASA regulations about informing Mission Control, and about freeze-drying the Commander's vomit as a sample for later analysis, doing neither. Their conversation was recorded unbeknownst to them, leading to a public reprimand, and kicked off the tension that later led to a strike. Illness, unfamiliarity and irritation led to the crew falling further behind with their work, whilst Mission Control tried to find more tasks for the crew in case they finished ahead of schedule. This micro-management and the unremitting pace demanded was perceived by the crew as unreasonable, with virtually no time off, and nary an uninterrupted meal. Later analysis showed that tasks in space took longer to complete than the same tasks on Earth. With mission controllers blaming the crew for their inefficiency and refusing to allow any meaningful downtime, the exhausted and angry crew went on strike six weeks into their mission, notifying mission control of an unscheduled day off, and switching off radio communications. Next day, the crew presented a list of demands. With no other option, Mission Control agreed to give the astronauts more flexibility over their time and work, with no interruptions during meals or in the evenings, allowing time to rest.

NASA learned to schedule downtime into all future missions to ensure astronauts mental wellbeing. However, none of the astronauts involved in the mutiny were allowed to travel in space again. Freeman (2000) suggests that the delay in resolving the situation was a "strike for rugged individualism", a supposedly typical and only American phenomenon. Kay (2014) on the other hand, on the libertarian communism site, sees the events of Skylab 4 as a class war in space, describing NASA's decision not to allow any of the crew back into space as "vindictive".

6.3.4 Mir June 1995

One well-documented rebellion occurred in June 1995 aboard Mir (Broad 1997) forcing ground controllers to reluctantly cancel the inspection of a solar-energy array. As Bond (2012) further reports in an obituary published 8 years after Strekalov's death, one of the cosmonauts involved, Gennadi M. Strekalov, had long experience both as a cosmonaut, and with various failures and disasters on the ground. Strekalov was the official flight engineer on Mir in 1990 and retired in 1995. However, he was that year selected to return to Mir with the first American to visit a Russian space station, during which sojourn he and Vladimir Dezhurov performed five spacewalks to prepare for the automated docking of the Spektr module. In one spacewalk they ran low on oxygen, having spent almost seven hours outside. The crew being increasingly tired and irritable, Strekalov finally snapped and started yelling, refusing to undertake an unplanned spacewalk on safety grounds. The Russian authorities finally gave way after two days, although managers in Moscow decided to punish and fine the cosmonauts 10,000 dollars for this transgression. Oberg, a space historian, reports (Bond 2012) that the cosmonauts went to arbitration to have the fine overturned: "they were fined, in effect, for mutiny. They had to go to court to get their flight bonuses." The mission ended successfully with the first docking between Mir and a US space shuttle.

6.3.5 Mir July 1997

Broad (1997) reports that a series of failures had occurred aboard the Mir space station; this caused Vasily Tsibliyev to rebel against Mission Control and refuse to perform certain tasks, including a repair mission. It was possible that Tsibliyev had real heart irregularities, but had also been shaken by a fire that had broken out on Feb 23, 1997, followed by a collision of a supply ship with Mir. The fire had blocked an escape route, whilst the collision forced the sealing off of the Spektr module, containing scientific experiments and Michael Foale's sleeping quarters. Doubting his own ability to work safely and productively, it was likely that as commander of the mission, Tsibliyev considered that a fresh crew should carry out repairs.

6.4 Astronaut Hadfield on Mutinies

Mutinies are clearly a subject of interest to the general public , attracting 35 correspondents discussing the possibility of a mutiny on the International Space Station during an Ask Me Anything session on reddit entitled *I Am Astronaut Chris Hadfield, currently orbiting planet Earth* in February 2013. Canadian astronaut Hadfield's reply was succinct:

[Q:] "Have you ever considered organising a mutiny, deorbiting the ISS and sailing the infinite void of space?"
[A:] "Those are several bad ideas: We're here for a pretty pure purpose, on behalf of everyone else. Keeps mutinies to a minimum."

Although the conversation amongst the participants rapidly degenerated in the way that it does on these multi-user forums, it was suggested that even if a mutiny did take place, the ISS would be unable to leave orbit because of a lack of propellant, as well as thermal design limitations. However, the central idea appears to be that the space station would move away from Earth as an independent entity. Note that Hadfield does not specify what the "pure purpose" is.

6.5 Cultural Differences

Differing cultural values are likely to be an important consideration in regulating and limiting acceptable behaviour or otherwise in settlements and habitats comprising crew from multiple cultures. Oberg (2000) describes a case of alleged sexual harassment as reported in the *Ottawa Citizen* during Sphinx-99, a 110-day analogue mock-space field experiment on observing group dynamics under varying conditions that took place through 1999 including New Year's eve seeing in the new millennium. Russian psychologists excused an incidence of alleged forcible French-kissing imposed on a Dr Judith Lapierre from Canada as normal Russian behaviour for such celebrations, emphasising that it was not an acceptable Russian norm to complain about such incidents. After complaints from the Canadian, Japanese and Austrian participants about this and incidents of fighting amongst the Russian participants, the incident was dealt with by providing locks for the tunnel between their module and that of the Russians'. This not only underlines the difference in cultural values and norms, but also draws attention to what is likely to be considered acceptable in cases leading to dissent, disagreement or rebellion in similar space environments. However, the Russian Space Agency also pointed out that Lapierre had violated a signed contract regulating protocols limiting personal behaviour imposing confidentiality, a case of breach of contract.

6.6 Regulating Behaviour on the ISS and in Space

The details of construction, operation and crew and command structures of the ISS are readily available on the Internet. The salient matter for our purposes is that the International Space Station (ISS) is not a military installation, and is operated under the auspices and aegis of the National Aeronautics and Space Administration (NASA), a civilian organization legislated for and funded by the Government of the United States of America (US or USA). It is operated multilaterally, with various components, services, resources and crew originating from other countries through the various and respective space agencies. As such, each space agency represents a sovereign nation-state on the ISS, with the exception of the European Space Agency (ESA), which is itself an institution with numerous co-operating signatories, each of authorised by their respective sovereign legislations.

6.6.1 Space Station (ISS) Intergovernmental Agreement (IGA)

The International Space Station is regulated by a multi-layered legal structure. At its core is the Intergovernmental Agreement specifying obligations and rights between the ISS partners, signed on January 28, 1998 by fifteen governments involved in the Space Station project: these being Canada, Japan, the Russian Federation, the United States, and eleven Member States of the European Space Agency (Belgium, Denmark, France, Germany, Italy, The Netherlands, Norway, Spain, Sweden, Switzerland and the United Kingdom). Article 1 states that this is “a long term international co-operative framework on the basis of genuine partnership, for the detailed design, development, operation, and utilisation of a permanently inhabited civil Space Station for peaceful purposes, in accordance with international law.” The IGA inherits and affirms the obligations of previous treaties and agreements that apply to the ISS, especially the Outer Space Treaty of 1967, the Rescue Agreement of 1968, the Liability Convention of 1972 and the Registration Convention of 1976. This of necessity implies that all laws be applicable under the jurisdiction of the originating state, further delimited in the Liability Convention. Criminal jurisdiction in particular becomes and remains the responsibility of the nation-state(s) involved in commissioning the component part used, in the case of a habitat module. It also reiterates that claims to national appropriation cannot be made over outer space or any portion of outer space.

Jurisdiction is again addressed in Section 2 of Article 5: “Pursuant to Article VIII of the Outer Space Treaty and Article II of the Registration Convention, each Partner shall retain jurisdiction and control over the elements it registers in accordance with paragraph 1 above and over personnel in or on the Space Station who are its nationals. The exercise of such jurisdiction and control shall be subject to any relevant provisions of this Agreement, the MoUs, and implementing arrangements, including relevant procedural mechanisms established therein". This is relevant for any actions resulting in mutiny or rebellion undertaken by personnel that may be considered contrary to safety.

The behaviour of crew selected by each partner to the IGA is subject to a Code of Conduct (Article 11, Section 2): "The Code of Conduct for the Space Station crew will be developed and approved by all the Partners in accordance with the individual Partner's internal procedures, and in accordance with the MoUs. A Partner must have approved the Code of Conduct before it provides Space Station crew. Each Partner, in exercising its right to provide crew, shall ensure that its crew members observe the Code of Conduct."

Liability is limited in this IGA through a cross-waiver; however, this cross-waiver of liability does not apply to "claims made by a natural person, his/her estate, survivors or subrogees (except when a subrogee is a Partner State) for bodily injury to, or other impairment of health of, or death of such natural person" or "claims for damage caused by wilful misconduct", in Article 16, Section. 2, sub-section d, paragraphs 2 and 3. Such claims may be relevant in cases of damage or injury caused by the actions of dissenting individuals.

Similarly, under Article 18, immigration is subject to the laws of the relevant Partner State. This authorizes the Partner State to exercise its jurisdiction such that a person can be deported or removed from its territory if any transgression and breach of law has occurred.

Article 22 extends the exercise of criminal jurisdiction of the Partner States over their respective nationals, and in cases of misconduct in orbit affecting life and safety, or causing damage, to the exercise of transnational criminal jurisdiction over non-nationals where agreed, including extradition based on the provisions of the IGA. However, the exercise of criminal jurisdiction by partner states does not override the provisions of the Code of Conduct agreed between Partner States. The Code in turn does not limit the application of Article 22, which is subject to each partner state's laws and regulations.

6.6.2 Memorandum of Understanding

There are 4 Memoranda of Understanding (MoUs) associated with and supplementing the Intergovernmental Agreement, one each between the USA and Russia, Japan, the ESA and Canada respectively. There are no MoUs between ESA, Roskosmos, CSA and JAXA because NASA is the designated manager of the ISS. These add conditions to the IGA. The Memoranda came into force at the same time as the IGA and are each identical, the only differences being the participants. Of relevance here are Art. 15 covering customs and immigration, and criminal jurisdiction, which are referred back to the IGA and the provisions made therein, whilst Art. 11 concerns setting up a Multilateral Crew Operations Panel (MCOP) and various other boards overseeing crew health, responsible for the specification of protocols and operations. In particular, Article 11 Section 8 directs a Space Station

Code of Conduct to be developed by the partner states, to be approved and implemented by the Multilateral Coordination Board. Note that immigration control and a restrictive territorial jurisdiction over crime is an oft-considered method of removing undesirables exercising civil disobedience, allowing the option of excluding or deporting those disobedients from one territory to another, usually that of their jurisdictional nationality and status.

6.6.3 ISS Crew Code of Conduct and Disciplinary Policy

Another legal layer of agreements implements and supplements the four MoUs further. Notable among them is the ISS Code of Conduct signed in 1998, setting out criminal jurisdiction, anti-harassment and certain other behavioural rules for ISS crewmembers.

According to Article 11, Section 8 of the MoUs discussed above: "The Space Station Code of Conduct will, inter alia: establish a clear chain of command on-orbit; clear relationship between ground and on-orbit management; and management hierarchy; set forth standards for work and activities in space, and, as appropriate, on the ground; establish responsibilities with respect to elements and equipment; set forth disciplinary regulations; establish physical and information security guidelines; and provide the Space Station Commander appropriate authority and responsibility, on behalf of all the partners, to enforce safety procedures and physical and information security procedures and crew rescue procedures for the Space Station."

Of direct relevance to the discussion about dissent at hand, it is, according to the Code of Conduct, the responsibility of the ISS Commander to "enforce procedures for the physical and information security of operations, to maintain order, to ensure crew safety, health and wellbeing, and to take all necessary precautions for the protection of ISS elements, equipment or payloads". It should be recognised that despite these general terms, these are the very elements, human or otherwise, that may be put at risk in the case of a mutiny or rebellion, depending on the actions of the disobedients on board. Apart from the ISS commander, the chain of authority aboard the ISS is determined on a mission basis, or on the changing situations as required, such as rescue or other. Although the chain of command includes the Flight Director on the ground, the ISS Commander is not obliged to implement the directives of the Flight Director, although this is generally accepted to be the case.

Use of equipment and goods is on the ISS is limited to the purposes of the ISS: this can imply that any misuse, outside the purposes of the ISS, should be seen as a disciplinary offence, or as an act of disobedience on the part of the crewmember.

There is also a Disciplinary Policy to be used to address violations of the Code of Conduct. The Disciplinary Regulations in the Code of Conduct note that violations may affect flight assignments, and that partner states can apply their own laws, regulations, policies and procedures, provided that this remains consistent with the relevant provisions of the IGA and the MoUs.

Neither the Code of Conduct, nor the Disciplinary Policy documents set out the nature of a violation of the Code of Conduct applicable to crew. It may be presumed that an act of disobedience would be considered a violation requiring discipline and punishment, but there appears to be no specific provision for this type of contingency.

6.6.4 *NASA Astronaut Code of Professional Conduct*

NASA has also issued a Code of Conduct for its Astronaut Corps, which requires comporting one's self to the highest standards of ethics and integrity. The detailed discussion of this document is not relevant to the present essay, except insofar as it strives to establish relationships of trust, emphasizing public duty, teamwork and the need to protect and balance the "best interests of our co-workers, families and NASA". To this end, the astronaut promises to "not take actions that put people or hardware at unreasonable risk." This last is a clear requirement to avoid harm or damage in any situation where the astronaut takes action in disagreement with authority.

6.7 Discussion

6.7.1 *Internet Forums*

As seen earlier, Internet users are also interested in questions of mutiny aboard the International Space Station. Space Exploration Stack Exchange is a question and answer site for spacecraft operators, scientists, engineers, and enthusiasts on which a general question about mutiny was posted in 10 October 2013 by user called2voyage (2013): "What would happen if a mutiny occurred on the International Space Station?", pointing out that as the ISS is a close quarter environment, and is not a military operation, a mutiny (however defined) could take place despite the careful selection and vetting of the crew involved, and requesting information about the laws and procedures that would govern such an event. His considered opinion is that neither the ISS Crew Code of Conduct nor Crew Disciplinary Policy helps in determining the answer in the case of refusal by a crewmember to comply with an order, nor does it help decide the conditions under which a rogue commanding officer would be replaced.

User KeithS provided an analysis of the situation and some possible solutions. These however do not take account of the legal jurisdictions in each module. Given that a mutineer cannot leave the station, he or she has very little privacy in which to "cool off", except when the ISS is not fully staffed. In the situation where the ISS Code of Conduct is violated, the ISS Commander can, under the ISS Crew Disciplinary Policy, be flexible and creative in dealing with the situation, perhaps locking the crewmember in a separate module, or subduing the crewmember through the use of drugs, brute force or the like, raising the possibility of damage to equipment, recognizing the lack of weaponry on board the ISS, and the difficulty of manœuvring in low gravity conditions. Further, whereas the ISS operations guide appears to consider “acute psychosis” and “suicidal thoughts or actions” to be medical emergencies, it is conceivable that a crewmember may revolt when suffering from these conditions, thus posing a threat to the entire crew and station, thereby requiring management and control through medication or restraint. There have apparently been no such cases reported to date. Apparently, Mission Control can also lock the crewmembers out of most station systems and run on autopilot, but KeithS does not provide any references for this. Note that this would be the equivalent of a siege situation, as measures may include withholding of critical supplies, including food, water and oxygen, whilst waiting out the rebellious crew. Non-cooperation includes the ground crew as a necessity, as anything else jeopardizes survival on the space habitat. A successful mutiny would therefore require the connivance of the ground staff, underlining and emphasising the inimical environment of space.

6.7.2 The Antarctic Treaty System

Given that Outer Space has been given the status of a Universal Commons, of being the common heritage of mankind, it is often assumed to have the same equivalent operating status as Antarctica. However, as pointed out by Winslow (n.d.) in his online Crime and Society: a comparative criminology tour of the world:

> Antarctica has no native population, and is inhabited primarily by scientists. Although Antarctica has no countries, several countries claim territories there. Neither the United States nor the former Union of Soviet Socialist Republics (USSR) has made a claim to Antarctica, and the claims to Antarctica made by Norway, France, Australia, New Zealand, Chile, Great Britain, and Argentina are not recognized.

During a Reddit driven Ask Me Anything (AMA) Q&A session (Afrosemite, 2012) entitled *I spent a year at McMurdo Station in Antarctica, AMA*, dont_-stop_me_smee asked about law enforcement in Antarctica:

> How do they deal with law enforcement down there? … Are there ANY laws in Antarctica? …You are a US citizen working at a federal facility. So all that comes with that still applies. Antarctica exists in kind of a neat, not in the USA but not anywhere else either thing. I heard it compared to being on a Navy ship in the middle of the ocean as a civilian. No jails or police, I think one of the guys during the winter had some sort of deputized authority. Didn't come up much

The Antarctic Treaty of 1959 does not appear to specify any legal structures, unlike the jurisdictional strictures of the Outer Space Treaties. From the general literature, it is apparent that many writers consider the Antarctic Treaty to be similar in application and use to the Outer Space Treaties. However, whereas jurisdiction is considered and specified in the Outer Space Treaties of 1967 and its four sequels, there appears to be no mention of jurisdiction in the actual Antarctic Treaty. This presumably reflects the claims of territoriality by the seven nations not including the then USSR and the USA over parts or all of Antarctica as mentioned above. As also mentioned above, it would appear, as in the jurisdictional authorities over their separate modules on the ISS, each country that operates a research station on Antarctica applies its own laws to its nationals in these research stations. Note however that most inhabitants of Antarctica are transient immigrant scientists subject to national jurisdiction. The international nature of Antarctic research stations and their jurisdictions is difficult to evaluate given the scarcity of strict applicable legal documentation, and the fact that each research station is operated by its respective country. The only actual international establishment appears to be a joint Franco-Italian station. Presumably nationals of another cooperating country that are visiting another country's installation are subject to the laws of that country, but may be subject to a trans-national application of their own country's laws. The details of Antarctica's regulations, although salient to the present essay as a contributory analogue, are beyond its scope and will be addressed in other work. However, conditions affecting psychological and social tensions in Antarctic installations are likely to be similar to those of space, deep sea and other similar isolated environments, and therefore may be considered indicative of possible motivations for dissent and rebellion (Harrison et al. 1991).

6.7.3 *Examples from Fiction*

Dissidence, disobedience and rebellion are common tropes in fiction, and in particular in the space sub-genre of science fiction, whether based in a military or civilian environment. Whereas there are numerous examples, too many to describe or cite in detail, two recent exemplars are noteworthy, *The Martian* (2015), a film based on the book by Andy Weir about a stranding on Mars, and the *Coyote* (2002 onwards) series of novels by Allen Steele about interstellar colonies, because they address some of the existing mythos current about space exploration, its enterprise and infrastructure.

6.7.3.1 The Martian [Film]

Piracy or Salvage

The protagonist in the film at one point describes the legal position regarding his intended appropriation of the *Pathfinder* probe left on Mars. It is not clear whether the craft is abandoned or not, but in appealing to Mars as being equivalent to international waters in terrestrial international maritime law, he concludes that he will become a pirate. This is an appeal to the concept of the universal commons and the common heritage of mankind referred to in the first Outer Space Treaty of 1967, which disallows the sovereign national appropriation of territory off Earth. Unfortunately, it is more likely, given that the craft remains under the jurisdiction of its originating state, it could be considered abandoned. Consequently, instead of the rather romantic figure of a rebellious pirate stealing and swashbuckling his way across Mars, he is more possibly a mundane salvager, acting according to the law, and therefore not even to be seen as lawless or breaking the law.

Mutiny and Rebellion

At the start of the film, a payload specialist, and therefore according to NASA not an astronaut, is presumed dead, and the plot relies on the mission's military command structure to take off from the surface of Mars during a sandstorm, with the pilot of the landing module deferring to the mission commander for orders. Later in the film, the Director of Flight Operations at the civilian agency NASA disobeys instructions from his immediate superior not to forward new flight instructions and details to the crew of the main spaceship. In doing so, he permits by omission a decision by the remaining crew to divert the spaceship in order to rescue the payload specialist stranded back on Mars. The decision to do so is suggested to be a democratic one: the collective decision is taken by a metaphorical show of hands. This raises the question of peer pressure and freedom of choice in a closed and hazardous environment, leading to a form of coercion which can therefore be deemed undemocratic. Two of the crew, the mission commander and the pilot, are part of a military hierarchy, and they are aware of the consequences if they mutiny, namely a court-martial. The remaining three aboard the spaceship are civilian payload specialists, and supposedly free to decide as they please. This draws a distinction between the concept of a mutiny, which is then a rebellion against the commands of an external military authority, or the direct disobeying of an order from the "captain" of a ship, and the absence of any civilian authority against whose decision the civilians act. This latter may not describe a mutiny, as even the mission commander is involved in the disobedience against the civilian agency, NASA.

Co-operation Between Agencies

At another point in the story, the Director of the Chinese National Space Agency is heard to suggest that the Americans be offered the use of their new and highly secret military booster to deliver aid to either Mars or to the rescuing craft. The offer cannot be made public, and must remain between space agencies, bypassing any approval from either the Chinese or the American governments. Not only does this suggest that there is somehow some special power to scientific and civilian cooperation which transcends traditional decision-making structures and institutions, but it is a clear act of disobedience in which the Chinese administrators of a government agency advocate the use of resources against all stated government policy on both sides of the Pacific. In the real world, NASA and the USA are set against Chinese involvement in the International Space Station, although there is no doubt some cooperation behind the scenes. Surely the Chinese government would also notice that one of their boosters is missing, or travelling in an unexpected direction, requiring a massive cover-up involving hundreds of disobeying individuals. The lack of cooperation between the USA and China is likely to be a consequence of laws regulating technology transfer under certain conditions.

6.7.3.2 Coyote [Novel]

Mutiny v. Hijack

Coyote is an unashamed story of the flight for freedom away from oppression, and is a well-researched work that purports in fictionalised form to describe the origin and growing pains of an extra-solar colony, relying on extrapolation from current proposals, and engineering and astronomical and astrophysical observations. It starts with replacing much of the crew and proposed colonizers of a colonizing starship with dissident intellectuals and those who oppose the liberal authoritarianism characteristic of the new regime that has established the new Republic of Alabama from the ruins of the older United States. Here, the captain of the ship is again involved, much as in *The Martian*. However, this is seen as a hijacking, a taking over a ship from its original crew, and not as a mutiny, despite rebelling against the government of the time, the putative owners of the ship. The novel and its sequels are also a clear appeal to the myth of the founding of America through dissent, referencing the flight of Protestants from England. The shuttles eventually land on the extra-solar moon around a planet belonging to Ursa Majoris B and are renamed the *Plymouth* and the *Mayflower,* although the starship remains the URSS *Alabama.*

Military or Civilian

In the scene where a new constitution is debated for the settlement, Steele draws attention to the separation between the military and civilian approaches to government by questioning the necessity of military ranks in a civilian economy:

> … if the purpose of an elected government was to put all members of the colony on an equal basis, formally retaining military rank would mean that "some citizens would be more equal than others. (Steele 2002, p. 227)

This can be contrasted with the joint structure described in *The Martian*, albeit the latter involves a different type of dissent and rebellion. It also suggests that there ought to be a separation of powers, the legislative from the executive at least, although enforcement sits astride both, and includes the judicial branch.

The Founding Myths

Above all, as stated previously, *Coyote* is a story of the supposed frontier spirit that built America, and the quest for the freedom that allows this, an ostensible denial and separation from the tyrannical and pre-existing authority left behind, leading to the establishment of an independent entity, a new and sovereign nation-state if you will.

Steele makes an interesting indirect suggestion in these works: that not only is rebellion required to found a truly free country, but that as the settlement grows, a necessary part of its evolution is another rebellion, that against the ideals or authority structures of the founding fathers. Regrettably, this is not an idea that he develops further, although the consequences may be significant if extrapolated in the light of the Outer Space Treaties of the Twentieth Century, allowing for settlements not under the aegis of an existing Earth sovereignty. In other words, a rebellion is seen as a necessary rite of passage, a birth or coming of age for the people.

6.7.4 Secession and Unilateral Declarations of Independence

Whereas mutiny is likely to lead a court-martial and consequent demotion, discharge, dismissal or removal from the hierarchy of the armed forces and its command structure, disobedience and dissent may lead to a renewal of the *status quo ante*. However, rebellion may lead to a unilateral declaration of independence, such as in the case of Rhodesia (now Zimbabwe) in the 1970s. Independence achieved, the new legislative body of the new country has the option of inheriting all treaties and continuing the decisions agreed to by previous governments exercising jurisdiction, or of abrogating those treaties. In the latter case, the effect of the various clauses and articles would be nullified, leading to a novel and sovereign entity, which nation-state would then be able to appropriate new territory off Earth. The rebellion that founds Coyote is a denial of belonging and citizenship; therefore there is no continuity of jurisdiction or of belonging to or within a sovereign entity. Then, a stateless person, a federal level administration, a non-governmental (but supra-governmental) organisation or an independent entity off-Earth not under the jurisdiction of a terrestrial nation-state may not be subject to the sovereign national appropriation limitations of the Outer Space Treaties, specifically that of 1967, allowing the establishment of new sovereignties in new territories. It is also consistent with the Hindu notion of migration to flee tyranny (Saksena 1941).

6.7.5 A Constitution Based on Deontology

The question then arises as to whether previous co-operating governmental agreements and disciplinary policies designed for modern outer space purposes may be the seed through which a new constitution may be proposed and propagated for the new nation. Is it possible to base a document for governance or a constitution for future artificial habitats on the IGA and deontology of astronauts? The deontology given above by NASA for their Astronauts' Corps does not allow for dissent in any but a very covert, implied form. Disobedience is disapproved of implicitly, as contrary to teamwork and the best interests of the people. This would then imply a somewhat rigid, authoritarian approach to a constitution, with an inflexibility of principle that would not be acceptable in the modern world, and would dismiss the deontological principles as contrary to individual conviction and conscience, and is perhaps a discussion best left for future work.

6.8 Conclusion

Examples over the decades of disobedience and dissent on space stations indicate that far from conviction or conscience, the astronauts on board the various artificial habitats have responded instead to overwork and the consequent need to take time off. Lesser concerns clearly provide better motivation than high ideals for immediate action. In contrast, the law demands adherence to some set of normative discipline, and this is enforced through the use of regulatory procedures. Punishment however is no more a legal necessity than is a legal definition of disobedience. This is achieved through the development and implementation of various agreements at a superior level, which then restrict normative behaviour through the use of a code of conduct or through a disciplinary policy. In all such cases, an essential element of the responsive behaviour to any given situation is the need, as opposed to the desire, to act against the demands of social hierarchies, which presumes a statement of these needs in the legal order applicable to individual behaviour. There is also a clear misunderstanding and consequent misapplication of the legal environment concerning dissidence in the popular imagination, although in both the real and the fictional world, there is an acceptance that rebellion and revolution may lead to a change "for the better", with the caveat that the "better" is a matter of individual perspective. This last then provides a justification for secession from existing jurisdictions and authorities, which may then allow for the establishment of extra-terrestrial settlements in accordance with international law and the treaties of Outer Space, which limit the presumed territorial appropriation off Earth.

Note that the analysis given above ignores any contribution from the Russian, Chinese or other space-faring nations, whilst also underlining that the dominant paradigms for these responses are primarily Western, and in that sense US–American, whilst the majority of nation-states and the various cultures on Earth with their differing attitudes to dissent remain unrepresented in the above discussion.

References

Afrosemite. (2012). *I spent a year at McMurdo Station in Antarctica, AMA* at http://www.topiama.com/r/155/i-spent-a-year-at-mcmurdo-station-in-antarctica. Accessed October 31, 2015.

Bond, P. (2012, August 15). Gennady Strekalov: Cosmonaut on near-fatal missions. *The Independent*, at http://www.independent.co.uk/news/obituaries/gennady-strekalov-16219.html. Accessed October 31, 2015.

Broad. (1997). On edge in outer space it has happened before. *New York Times*, at http://www.nytimes.com/1997/07/16/world/on-edge-in-outer-space-it-has-happened-before.html. Accessed October 31, 2015.

Brownlee, K. (2012). *Conscience and conviction: The case for civil disobedience*. Oxford: Oxford University Press.

called2voyage. (2013). What would happen if a mutiny occurred on the International Space Station? *Space Exploration Stack Exchange*, at http://space.stackexchange.com/questions/2372/what-would-happen-if-a-mutiny-occurred-on-the-international-space-station. Accessed October 31, 2015.

Donovan, J. M., & Anderson, H. E, I. I. I. (2003). *Anthropology and law*. New York: Berghahn Books.

Freeman, M. (2000). *Challenges of human space exploration*. Chichester: Springer Praxis.

Fuentes, J. (2015, March 13). *5 True outer space Mishaps* at http://www.fromquarkstoquasars.com/5-hilarious-mishaps-outer-space/. Accessed October 31, 2015.

Goh, G. M. (2007). *Dispute settlement in international space law*. Leiden: Martinus Nijhoff Publishers.

Hadfield, C. & Participants. (2013). *I Am Astronaut Chris Hadfield, currently orbiting planet Earth* at https://www.reddit.com/r/IAmA/comments/18pik4/i_am_astronaut_chris_hadfield_currently_orbiting_planet_earth. Accessed October 31, 2015.

Harrison, A. A., Clearwater, Y. A., & McKay, C. P. (Eds.). (1991). *From Antarctica to outer space*. New York: Springer.

Kay, S. (2014, November 8). *Class war in space—The Skylab 4 mutiny* at https://libcom.org/history/1973-skylab-4-mutiny. Accessed October 31, 2015.

Oberg, J. (2002). Violence and sexual assault in "Space". *James Oberg's Pioneering Space* at http://www.jamesoberg.com/04142000assualt_rus.html. Accessed October 31, 2015.

Right of Revolution, Wikipedia, accessed October 31, 2015.

Saksena, K. P. (1941). *Hindu law and jurisprudence with mimansa rules of interpretation*. Allahabad: Rai Sahib Ram Dayal Agarwal.

Steele, A. (2002). *Coyote*. Orbit.

Teitel, A. S. (2013). *Skylab 4 Rang in the New Year with Mutiny in Orbit* at http://motherboard.vice.com/blog/ringing-in-the-new-year-with-mutiny-in-orbit. Accessed October 31, 2015.

The Martian [film]. (2015). dir. Ridley Scott, based on the book by Andy Weir, self published 2009.

Vitello. (2014). William Pogue Astronaut who staged a strike in space dies at 84. *New York Times* at http://www.nytimes.com/2014/03/11/science/space/william-r-pogue-astronaut-who-flew-longest-skylab-mission-is-dead-at-84.html. Accessed October 31, 2015.

Winslow, R. (n.d.). *Crime and Society: a comparative criminology tour of the world* at http://www-rohan.sdsu.edu/faculty/rwinslow/antarctica.html. Accessed October 31, 2015.

Treaties, Agreements and Codes

Astronaut Code of Professional Responsibility (2008) NASA.

Code of Conduct for the International Space Station Crew (2000) Multilateral Crew Operations Panel (MCOP).

Disciplinary Policy for International Space Station Crew (2000) Multilateral Crew Operations Panel (MCOP).
International Space Station—*Intergovernmental Agreement* (1998).
International Space Station—*Memorandum of Understanding (1998) NASA-ESA* 1998 NASA.
International Space Station—*Memorandum of Understanding (1998) US-Canada* 1998 NASA.
International Space Station—*Memorandum of Understanding (1998) US-Japan* 1998 NASA.
International Space Station—*Memorandum of Understanding (1998) US-Russia* 1998 NASA.

Chapter 7
Space Terrorism—A New Environment; New Causes

John R. Cain

Abstract This chapter examines terrorism in a new environment namely space. There is an introduction to the issues relating to space terrorism from both a terrestrial and extra-terrestrial perspective. This is followed by a brief history of terrorism and a discussion on how terrorism is to be defined on Earth and in space. The types of terrestrial and extra-terrestrial terrorism are detailed and examples given of both. There is a discussion on the measures that terrorists may use to achieve their objectives and an outline of the countermeasures that could be used to prevent terrorist acts in particular as they relate to space. The chapter then examines specific measures to prevent acts of space terrorism including likely punishments and the possible need for an extra-terrestrial Human Rights Charter to prevent acts of space terrorism. Alien terrorism is then discussed and the affects alien encounters could have on relations between species.

Keywords Space terrorism · History · Definition · Types · Countermeasures · Prevention · Human Rights Charter · Alien terrorism

7.1 Introduction

Terrorism is not a modern phenomenon but has been in action in all societies throughout history as a means to generate fear in a population. However, it is only recently that it has become a force for major social change, especially in modern Western liberal societies (Caruso 2014). This is because the terrorist groups including the individual members of such groups have access to modern weaponry and also a strong belief in their cause. When used, this combination has been shown to inflict major damage for example, by the London bombings in 2005 that resulted

J.R. Cain (✉)
GeoFind Consultancy, Hookstone Chase, Harrogate, North Yorkshire, UK
e-mail: johncain6@hotmail.com

C.S. Cockell (ed.), *Dissent, Revolution and Liberty Beyond Earth*,
Space and Society, DOI 10.1007/978-3-319-29349-3_7

in many civilian casualties. Unless there are significant economic, political and cultural changes in how populations respond to social change, then acts of terrorism will continue. In space, where societies will live in isolation it is likely that the settlements and spacecraft populations will be at an increased risk from attacks from those who have major grievances. It is therefore necessary that as man begins to explore and colonise space, that terrorism in all its manifestations is understood including the reasons why it is used by individual(s) or group terrorists to further their causes.

There has been a rapid increase in the use of individual communication systems within the past two decades (Chen and MacCauley 2011). Terrorists, therefore, have had access to cyber space technology and a potentially wide audience to utilise various websites such as Face-book and Twitter and to try and radicalise people who may have terrorist tendencies. Within seconds, such technology can be used to influence people all over the world by having access to political events as they happen including film of the events. This knowledge can be used to quickly spread radicalism amongst susceptible individuals and population groupings. For example, in the recent so-called Arab spring, cyber technology was used by dissidents to encourage uprisings to inflict maximum pressure on the ruling classes. And susceptible Muslim men and women in the United Kingdom have become radicalised by members of the Islamic State (IS) using specific websites such as You-tube to encourage recruitment to their cause and to travel and fight for IS causes in Syria and elsewhere.

Such events may seem remote from future manned space exploration and the establishment of settlements on the Moon and Mars. However, in space, the methods that the terrorists will use to create fear in populations, combined with the use of cyber space communications to promote their causes amongst settlement societies, will be similar to those as used on Earth. The nature of terrorism and the types of terrorism in all their forms need to be understood and countermeasures developed to prevent a major catastrophe in space.

The fear of terrorism may be misplaced because it only affects a very small percentage of a societies' population on Earth and these are mainly civilians (Grenshaw 1995). This will not be the case in space where an attack on a settlement will wreak havoc on all sectors of society. This climate of fear may cause governments to restrict the freedom and liberty of the settlement population for example, by restricting the use of specific internet sites, by restricting travel and by limiting the worship of a specific religion. Sen (2010) argues that there may be a reasonable case for placing elimination of the fear of terrorism within the concerns of human rights even if the fears are stronger than probability statistics would justify for example, there was a strong fear of further terrorist attacks throughout Turkey following the Ankara bombings in 2015.

7.2 Space Terrorism-Setting the Scene

The causes of space terrorism may not be easy to identify as a number of circumstances may lead to an attack. For example, Sen (2010) has stated that the consequences of an intervention by a Government either militarily or by applying sanctions to limit terrorist actions could lead to an upsurge in terrorism instead of reducing the dangers from attacks. This concept could also apply extra-terrestrially. If an Earth governing body wished to impose taxes on the supply of minerals necessary for ensuring the economic survival of a lunar settlement, then retaliation could occur against the Earth governing body by attacking its interests on Earth or against those settlements supporting the action. How the inflicted lunar settlement acted in response to the demands for more taxation and the subsequent retaliatory consequences, could be interpreted as justice for one settlement but a terrorist act by others i.e. space terrorism.

During the establishment of settlements on the Moon and Mars, the implementation of an interstellar social contract including a code of justice, freedom and liberty to peacefully settle disputes between settlements could reduce the need for terrorist acts over time. But Pogge (2001) has noted that if one party (e.g. a terrorist or religious group) tried to reduce tension by peaceful means such as the implementation of a social contract with duties for all, then the resulting impartiality and universalist intentions could create serious social problems. In a closed, self-contained Martian settlement, competition and antagonism with other settlements on the planet could result in the use of terrorist acts to ensure that outsiders are kept out and the settlement can develop in its own way without interference from others i.e. as a totalitarian state using state sponsored terrorism to further its causes (Rawls 1993). The insularity of the settlement's society may reflect the state of "happiness" and "well being" of the society with a desire not to promote the values. However, if a settlement is economically poor and depressed because of the imposition of perceived policies from Earth or from neighbouring more powerful settlements then this could lead to acts of terrorism to obtain a "better deal" for their citizens.

Goodin and Jackson (2007) have like Sen (2010) indicated that there can be reasonable causes for replacing the elimination of the fear of terrorism within the concerns of universal human rights even if the fears are stronger than probability statistics would justify. This sentiment may not be reflected in the space environment because many settlements may seek their own agendas to ensure their survival and to promote their own specific causes to ensure the environmental sustainability of their society. The widening of universal human rights as they apply extra-terrestrially may not therefore gain wide approval and will need to be revised to encompass those situations if they are to succeed. Will the settlements that are isolated because of their distance from Earth and neighbours be able to provide the necessary social, political and economic conditions to provide security and prevent space terrorism? If not, could this lead to terrorist acts on an interstellar scale? How will the imposition of human rights (and possible alien rights) be challenged and by whom, even if these rights are realised (Chatterjee 2008).

The interdependence of a settlement's population, especially if located at a distance from Earth, could result in a sense of isolation, and impact on their sense of injustice if they feel they are being discriminated against by another outside party. A perceived threat to a settlement could lead to acts of terrorism against others to ensure it's survival. Furthermore, discontentment in a settlement society, in particular if it is under the major cultural influence of another more powerful settlement, could spread quickly. It is therefore important that all those settlements that are under the cultural and economic influence of more powerful settlements be involved in the decision making processes to reduce the potential for alienation and the threat of violent action (Held 1995).

The examples of possible causes of space terrorism discussed above, question whether the implementation of an effective Human Rights Charter could apply both terrestrially and extra-terrestrially over time to reduce acts of terrorism or whether there would need to be limitations on its use. But to understand these issues, the definition of terrorism needs to be clarified and the types of terrorism that have been used historically on Earth to achieve specific goals identified? Will these types of terrorism apply extra-terrestrially where societies will mainly be isolated and struggle to survive in hostile environments? What measures will terrorists use in space to create fear in populations? What countermeasures, including the threat of punishment and the use of sophisticated surveillance techniques, will the ruling space governments use to combat terrorists and to prevent radicalism in all its forms?. Will contact with aliens result in new forms of alien terrorism? These questions will be addressed below. Burke (1986) wrote of the French Revolutionaries, that a belligerent ideology, bent on eradicating all opposition to its claims can be particularly aggressive. This view is likely to apply in similar circumstances extra-terrestrially especially if there is seen to be no justice for one or more settlements with a political, economic, scientific or social grievance that cannot be solved peacefully.

7.3 A Brief History of Terrorism

Acts of terrorism against states or specific target groups to further one or more causes have been used since the establishment of societies for example, the violent retaliation by the Jews against the Romans in the 1st century in response to religious persecution (Chaliand 2007). However, as societies developed and became more liberal, the practice of terrorism to promote particular grievances became targeted towards the ruling classes in a society. In the 19th century there was the French Revolution with the aristocratic fear of being beheaded by the revolutionaries. Later in that century there were anarchists, communists and socialists who wanted to destabilise society and create a so-called utopia on Earth. The 20th century saw acts of terrorism occur on an even grander scale. Both Stalin in Russia and Pol Pot in Cambodia used the starvation of peasants, the fear of imprisonment and random killings to further their murderous causes (Schmid 1994). Terrorism has flourished in the 21st century, in particular with the rise of Islamic extremism (e.g. Al Qaeda and IS) against the Western liberal democracies including the USA and the United Kingdom to spread their beliefs using violent and brutal means (Hoffman 2006). The conquest of space later in this century will result in terrorist attacks as specific governing groups in settlement societies compete against each other and struggle to survive in a hostile environment. Such acts will then become part of terrorist history.

7.4 The Definition of Terrorism

At present, what constitutes an act of terrorism on Earth and those likely to occur in space? What is the definition of a terrorist group (e.g. Hamas, IS, Maoists) and what distinguishes the actions of that group from another group especially if they both want to promote their causes by the use of violent acts and to generate fear in a society? And how far should a terrestrial or a future, extra-terrestrial government acting on its own or on behalf of other parties pursue those causing terrorist acts without infringing their human rights and interfering in the internal affairs of other states (Gellner 1983)?

The definition of terrorism has been debated by politicians, by academics and others for many years but there is still no universal consensus as to how it should be defined. This lack of an effective definition has been an obstacle to meaningful discussion and the subsequent development of suitable countermeasures targeted at specific individuals or groups. If there are difficulties in defining terrorism terrestrially because of the wide range and type of terrorist causes, it may be simpler in the early stages of space exploration because the causes of terrorism and those responsible should be easier to identify. But even though the causes of extra-terrestrial terrorism may be fewer, the actions to achieve their goals may be more violent in scale e.g. the destruction of a settlement's life support systems could destroy the settlement. And the vast distances of space may restrict communications and thereby limit an effective response. Due to their insular nature, those living and working in space stations and those travelling in large world spaceships are likely to generate a specific self-contained form of terrorism that will need to be categorised (Hein et al. 2012).

Acts of terrorism could be defined as peacetime equivalents of war crimes (United Nations Crime Branch 1992). However, this definition is fraught with complications because a definition would need to be found for war crimes in particular if the "terrorist acts" were carried out in space where there have been no terrestrial war equivalents. As early as 1937, the League of Nations Convention defined terrorism as "All criminal acts directed against a State and intended or calculated to create a state of terror in the minds of particular persons or a group of persons or the general public" This definition is very broad sweeping and this is why it probably never gained acceptance.

In 1988, a group of academics (Schmid and Jongman 1988) produced a consensus definition of terrorism:

> Terrorism is an anxiety-inspiring method of repeated violent actions employed by (semi-) clandestine individual, group or state actors for idiosyncratic, criminal or political reasons whereby - in contrast to assassination – the direct targets of violence are not the main targets. The immediate human victims of violence are generally chosen randomly (targets of opportunity) or selectively (representative or symbolic targets) from a target population and serve as message generations. Threat and violence based communication processes between terrorist (organisation),(imperilled) victims, and main targets are used to manipulate the main target (audiences), turning it into a target of terror, a target of demands, or a target of attention depending on whether intimidation, coercion or propaganda is primarily sought.

This is a wide-sweeping definition of terrorism that tries to incorporate all acts of terrorism within its umbrella. For this reason this academic consensus definition was later revised due to its complexity (Schmid 2011).

In a study of the definition of terrorism, Bruce (2013) has stated that because many groups such as academics and governments have a vested interest in terrorism and pursue different agendas, that it will be unlikely to produce a common definition of terrorism. He is probably correct. As Man conquers space their may be additional vested interest groups such as astronauts and space entrepreneurs with their own agendas and this will make a broad definition even more confusing. However, a definition of terrorism as applied to space needs to be attempted and could simply be:

> Space terrorism is an act of violence by one or more individuals or groups to prevent the development of a space settlement(s) and/or their aims including those of a spaceship or space station during Man's exploration of space.

7.5 The Types of Terrorism on Earth and Likely to Occur in Space

The types of terrorism that occur on Earth and the causes that they are related too, have developed over many centuries in response to changing political, economic, social and religious circumstances. If there are no major cultural changes resulting in a universal condemnation against terrorism in all its forms, it is expected that there will be a similar pattern of terrorist development extra-terrestrially albeit on a different scale. The main differences between terrestrial and extra-terrestrial acts of terrorism are likely to be:

- in the scale of the acts of violence (e.g. destroying a large space station may result in many casualties);
- on the sophistication of the acts (e.g. the use of advanced weapon technologies with pin-point accuracy to destroy specific targets);

Table 7.1 The types of terrorism, their definition and examples on Earth and in space

Types of terrorism	Definition	Examples on Earth	Examples in space
1. State terrorism	Use of terror by Governments to control its population	Turkey against the Kurds	An Earth Government to control Moon colonies
2. Religious terrorism	Motivated by religious ideologies and grievances	Al Qaeda, IS	Rise of new religion in space and terrorist acts to promote
3. Right wing terrorism	Use of violent militias and gangs to gain power	Neo-nazi groups in Germany	The promotion of an elite force of astronauts for control
4. Left wing terrorism	Seek to overthrow democratic Governments	Red Path in Peru	High tech groups to overthrow chosen government
5. Pathological terrorism	Use of terror by individuals for pleasure	School shootings in USA	Blowing up of planetary rover with passengers
6. Separatist terrorism	Seek to cause havoc in a country to establish a new state	ETA in Spain	Violence to ensure independence of one or more settlements
7. Narco terrorism	Organisations that fund the selling of drugs and use violence	Cartels in Mexico	Spreading of new drugs in one or more settlements or world ship
8. Eco terrorism	Violence in the interests of the environment	Environmental groups	To protect life discovered on a planet
9. Cyber terrorism	Use IT to attack Governments or civilians	Intercepting drones and firing missiles	Threat of use of IT to close life support systems
10. Science terrorism	Withholding scientific data resulting in death	Tobacco companies and smoking	Failing to use synthetic biology to produce food in settlement

- on the limitations of the physical environment (e.g. the distances and isolation between settlements may deter ground attacks); and
- on the response to the acts (e.g. Earth involvement in the use of countermeasures may act as a deterrent).

The mass slaughter resulting from an act of space terrorism could create a new type of terrorism in that whole populations could be destroyed by one act of violence. There would be little help for those that survived or were injured. For example, Crawford and Baxter (2014) have indicated that a manned Mars transport using an inertial confinement fusion propulsion system (Orth 1990) could be used to destroy a city on Earth or a large settlement on Mars just by crashing onto it. This would be a suicide mission but one that could be contemplated.

At present, there are a number of types of terrestrial terrorism that can be identified for example, state terrorism, religious terrorism and cyber terrorism (Grothaus 2014). These terrorism types as applied on Earth can also be used to categorise possible space terrorism. Table 7.1 lists the types of terrorism identified so far and provides a brief definition of each type with examples terrestrially and extra-terrestrially.

It is difficult to predict the type(s) of terrorism that will be the most prevalent in space. However, because the development and application of science will be necessary to survive in the hostile environments of space, the use of science terrorism (i.e. withholding and/or not applying scientific knowledge resulting in many deaths) may be the main means used to achieve the goals of a small but powerful group of scientists/technocrats who wish to gain power and promote their causes. Furthermore, the need for a specific and elite group of highly skilled astronauts (Cain 2014) with the influence that this will ensure, may see a rise in "right wing" terrorism (from the astronaut class) against a potentially "left wing" group similar to the Samurai as described by H G Wells in *A Modern Utopia* (2005), to achieve their aims.

7.6 The Means of Achieving Terrorist Acts in Space

7.6.1 Background

Any terrorist group whether on Earth or in space, will need access to the means of causing violence. Apart from the provision of weapons and the knowledge on how to use them, it will be necessary to have the finance to purchase the weapons and the knowledge to promote propaganda (e.g. by the use of cyber communications) and to have influential backers for their causes. State and religious sponsored acts of terrorism will have the most generous backers and the greatest support to pursue their aims.

If a terrorist group has access to the means of carrying out an attack and needs to ensure its success, then those chosen to commit the act of terrorism will need to be persons unknown as terrorists by the authorities. For these reasons, most terrorist targets extra-terrestrially will likely be chosen by sympathetic groups on Earth who cannot be identified and will then travel to the Moon or Mars for example, to carry out their acts of violence.

Because of the closed nature of a space settlement and the need for effective life support systems at all times in particular the provision of oxygen and water, any terrorist act against such facilities could have devastating consequences. Furthermore, a strike against a long-life fission power system used on the Moon or other outposts to generate electricity would also cause mayhem and generate unrest and fear in the population even if a reserve power supply was available (Schriener and El-Genk 2013).

In order to reduce the risks of a terrorist attack, it may therefore be necessary to stringently monitor all those people travelling on a spaceship and those entering and leaving a settlement. But the use of an unidentified "home grown" radicalised suicide attacker, as a "weapon", to undertake a terrorist attack, (for example, to destroy an oxygen manufacturing facility on the Moon or Mars), would have major repercussions throughout space and those on Earth.

To ensure the success of an attack, the individual terrorist or group terrorists would need an effective tracking system to monitor their target(s) over a period of time before striking. In a settlement or travelling in a large starship, this would include close and timely observations of the target population prior to undertaking any action.

7.6.2 Destruction of Satellites as a Terrorist Means to Achieve Success

A terrorist group could cause major disruption to a number of space sectors by the direct destruction of one or more satellites operating in the sectors using laser technology (Dinerman 2006). If inter-spatial communications between settlements and Earth depended on a network of satellite communication systems, then the destruction of a satellite(s) would isolate and jeopardise the survival of a target population.

Alternatively, the terrorist could destroy or severely limit the workings of a satellite(s) by preventing the service and maintenance of such technology e.g. by kidnapping the engineers, by destroying the satellite monitoring systems on Earth, Moon or Mars or by destroying the anti space debris prevention equipment. Once destroyed it would take time to repair or replace the satellite(s) during which time there would be much hardship by those affected. Maybe the threat of destruction of a satellite(s) by a terrorist group would be sufficient to meet their demands?

7.6.3 Terrorist Response to Social and Cultural Changes

As the influence of Earth's governing classes weakened in controlling the politics and economics of the space settlements, then separatist groups could see this as an opportunity to spread their propaganda by violent means. In such circumstances, the use of effective communication systems would be vital to indoctrinate and radicalise susceptible individuals on Earth and in space. To reflect the profound social, economic and cultural changes occurring throughout space and as various groups struggled to gain power, there would likely be an increase in terrorist activities in the settlements and also on Earth. Following the breakdown of a settlement's authority, hegemony and legitimacy, a power vacuum would occur that terrorist groups would seek to exploit (Hobsbawm 2007).

In such social and economic crises, certain actions undertaken by Earth and the settlement governing bodies could instigate a situation that would lead to acts of terrorism. If a large conglomerate on Earth restricted the use of specific metals in a settlement and/or increased the cost of oxygen, both items vital for survival, then this would lead to hardship and restrictions on the freedom of movement within a settlement to limit use. The implementation of appropriate action by affected settlements to reverse the decisions would need to be carefully evaluated because any action could be interpreted as a terrorist act and countermeasures instigated by the enforcing authority leading to possible conflict.

During a long-term space mission in a starship or during living and working in a large space station there may be those that object to the system of administration. These "terrorists" could decide to destroy or threaten to destroy one or more aspects of a safety system that ensures sustainability in one or more sectors of the craft (Dulo 2014). But in such spacecraft with limited environmental space, any severe damage to the vessel would be suicidal. This may not matter to a terrorist group if the destruction was part of a broader strategy to achieve their aims.

7.7 Counter Measures Against Terrorists in Space

It is necessary to have effective terrorist countermeasures in place to prevent or to be able to successfully retaliate against a terrorist attack that may cause devastation to property and result in physical and psychological damage to persons. The countermeasures used will also need to protect and maintain the rule of law both on Earth and in space, whatever the political system. At present, it is only national governments that have the resources necessary to successfully and effectively fight and deter terrorism. In the future, it will be the governing authority in a settlement working in collaboration with Earth that will take on this responsibility.

Three effective measures to counter terrestrial terrorism have been developed. These include the Political Pathway Model, the Military Solution or War Model and the Criminal Justice Model. All of these models either alone or in combination have been tried (Hoffman 2002) by many governments but they have had only limited success. If the three Models are adapted for use in space, it is expected that they will also have limited success unless efficient and effective profiling of potential terrorists can be made and the most appropriate Model can then be used to target the terrorists based on the profiling.

7.7.1 The Political Pathway Model

This Model addresses the motives of terrorists politically and diplomatically and strives to arrive at a means of resolving the issues peacefully (Pedahzw and Ranstorp 2001). However, as a policy used by Governments, it has not been successful because of disagreements as to who the terrorists are and who is supporting them. As part of the use of this Model, Remuss (2009) has argued that the European Union (EU) should have its own measures to counteract terrorism as this would strengthen national capabilities within the Union and promote international partnerships (Council of the European Union 2005). But inside the EU the perception of a threat will differ amongst nations due to history, national demographics and national characteristics and this could reduce the effectiveness of any counter-measures. If acts of terrorism occur against space settlements with differing economic and political systems then such a parochial political approach may be difficult to implement (Hertzfeld 2007).

7.7.2 *The Military Solution or War Model*

The use of the military to combat terrorism has had a limited success terrestrially; it utilises the armed services to directly challenge a terrorist or a terrorist group (e.g. IS in Syria), or by the use of infiltration to gain knowledge of attacks (e.g. Irish Republican Army in Northern Ireland). Extra-terrestrially, the use of the military will need to involve the use of sophisticated weaponry to prevent or control the scale of a potential terrorist attack on satellites, space settlements or inside a spacecraft. This will be costly. To be successful in their attacks, the terrorists would also need access to costly space based weaponry, the acquisition of which could be easily monitored by the appropriate military authorities using knowledge gained by infiltration. Similarly, in spacecraft and space stations, terrorists would find it difficult to smuggle weapon systems aboard especially if effective detection equipment was used. Only those terrorists with access to sufficient finance and the ability to build their own weaponry and delivery systems would pose a threat to selected targets on Earth and in space (Peoples 2008).

The use of the military could be an effective option to prevent terrorists jamming satellites orbiting the Earth (Dinerman 2006). If given sufficient resources, surveillance cyber technology systems could be used to prevent terrorist:

- use of lasers to directly destroy a space-station;
- direct attacks on rocket launch pads/space ports; and
- cyber-security penetration linked to military establishments.

Further counter measures that the military could use to combat terrorism could include the use of tracer systems and associated anti-jamming devices and anti-missile systems in particular against direct military attack. However, the use of such counter measures could be overcome by a large well financed terrorist group using advanced cyber surveillance systems.

Terrorists could attack or threaten to attack the interests of large conglomerates involved in space exploration and who would be severely damaged as a result. The use of the military could be used to combat such attacks especially if they were direct. But because of the high cost of space exploration and the potential loss of life following a determined and successful terrorist strike, the scale and magnitude of such an attack would be enormous (Thomson 2006).

7.7.3 The Criminal Justice Model

The Criminal Justice Model involves the investigation of terrorist acts using the police and judiciary system of the country that was affected by the terrorist act and this includes any punishment given. A similar Model based on that used by countries on Earth could be implemented to protect the rights of individuals in settlement societies (Listner 2005). However, because of the logistics of operating within the vast distances between settlements and Earth, it could be difficult to track and capture the terrorists as they would find unexplored sectors of space in which to hide and evade the law (Newlove and Erickson 2011). It would therefore entail more resource to bring them to justice in particular if they were being sheltered by sympathisers living in a settlement or space station.

An extra-terrestrial police or quasi military force would need to be established, possibly in the early stages of space exploration, to track down terrorists and bring them to justice. The anti-terrorist force would need to be universally accepted and represented by all parties in space and on Earth otherwise it would not function effectively. Until such an enforcement body was established, there would need to be an effective justice system that represented the country or settlement(s) under attack. But this could cause mistrust between affected parties especially if specific anti-terrorist legislation didn't apply to all those involved. It is important that policy makers discuss these issues now while the conquest of space is in its infancy (Crelinsten and Schmid 1993).

It is not possible to predict the forms of terrorism that may occur in space or on the measures that terrorist groups will find most effective to further their causes. But to combat terrorists over the vast expanses of space, the implementation of a combination of the three Models may be the best counter measure to deal with terrorist grievances. Any policy that is used will require an interstellar court of independent, impartially minded individuals with a wide remit to distinguish the root causes of the terrorists and to implement harsh penalties for guilty individuals such as banishment to dedicated regions of outer space for time specific periods as determined by the court. But as life expectancy increases due to advances in science, then this punishment option may have moral and ethical implications for example, how long should a human being spend in isolation on a hostile world. Furthermore, because of history, some settlements could deviate from the established rule of law and introduce their own special ant-terrorist legislation that may be more lenient to the causes of the terrorists.

7.7.4 Countermeasure Strategies

To succeed in combating terrorists using one or more of the Models described, the political governing classes will need to develop an extra-terrestrial terrorist countermeasure strategy to include all the specific anti-terrorist organisations that represent the settlements in a specific sector of space. The strategy will need to incorporate the best points of each Model to be effective as each only addresses part of the problem.

The authorities will need to respond to the fall-out of a terrorist attack and to apply emergency procedures. Effective protective measures will depend to some extent on a number of factors such as the availability of space dependency links via positive communications within an interstellar range and at trying to preserve the social functions of a settlement(s) e.g. by ensuring life support systems still functioned. The object of any survivability strategy would be to make targets more difficult to find, hit or kill. This may be hard to achieve because of the inter-dependency of those living and working in space and where major damage could not only affect a settlement but also those in a space-station who may have a symbiotic relationship with a settlement. Any extra-terrestrial counter-terrorism measures will therefore need to ensure close cooperation with Earth governments via local communication networks in particular in the early days of space exploration when terrorist groups may want to stage a major impact attack to promote their cause.

A start to establish a space terrorism countermeasure strategy has been made by the European Union to address the issue of space terrorism (Rathgeber 2008). The result of negotiations may be the establishment of a Charter similar to that of the UN Charter for Human Rights (1948) but incorporating the rights of extra-terrestrial populations. These rights may need to be more draconian in approach than if applied to Earth only, mainly because of the potential damage of a successful attack. Some of the draconian measures may include the use of genetic screening to identify and investigate those individuals whose psychological characteristics may cause them to act violently in certain situations and by the use of widespread surveillance to incorporate all those living and working in a settlement. But such restrictions on freedom and liberty may be a price that the space explorers are not willing to pay! Table 7.2 lists the measures that could form part of a future Charter to prevent acts of terrorism in space.

Rawls (1993) has indicated that it may be difficult to establish a world or extra-terrestrial federal body to deal with terrorism in all its forms. Terrorists are highly unlikely to follow the judgements of a world body and it may be sceptical to demand the setting up of such a body and promoting a social contract that all may follow (Pogge 2001). Each country, society and settlement will have parochial beliefs that they would like to promote and this could prevent the working of terrestrial/extra-terrestrial courts to deal with those effects of terrorist actions. (Nagel 1970). Maybe local justice restricted to one or more settlements on the Moon or Mars may be sufficient to limit the actions of an individual terrorist or a terrorist group and to punish those involved. But there may be naivety in such a parochial approach especially where the decisions made by one settlement governing body may encroach on those of another via. cross-settlement and geographical borders (Sen 2010). In such as case, the view of Rawls (1993) may apply:

Table 7.2 Measures to incorporate into a Human Rights Charter to prevent acts of terrorism in space

Measures to prevent acts of terrorism in space
1. Check political background of those leaving Earth
2. Space settlements and space stations to monitor those with views that are counter to those of their governing body e.g. politically, scientifically
3. Establish a space/Earth forum to discuss and identify grievances in one or more groups and determine solutions
4. Enforce strict punishments for those contemplating or carrying out acts of terrorism extra-terrestrially. This may include punishing those scientists who are not willing to use their expertise to ensure the long-term survival of a settlement
5. Implement controls on individual communications systems in particularly if they are being used to radicalise or promote violent causes. This will also include limiting the use of subversive literature to further a cause
6. At all levels of education, inform the students on how issues that could have violent undertones should be resolved peacefully
7. Implement strict licensing laws on weapons exported from Earth and stringent regulations on the use of such weapons in space. Software and the products of synthetic biology should be tagged and monitored to ensure that they are used for lawful means
8. Use genetic screening (e.g. exposomics) to identify and investigate those psychologically programmed to act violently in certain situations (Cain 2014)
9. Form an extra-terrestrial legislative justice body to oversee the above measures and to ensure that any punishments given to terrorists are proportional to the level of the crime

> I assume that the basic structure is that of a closed society: that is, we are to regard it as self-contained and as having no relations with other societies. That a society is closed is a considerable abstraction justified only because it enables us to focus on certain questions from distracting details.

Furthermore, it may be that the concept of the "impartial spectator" (e.g. human rights campaigners, scientists) and as defined by Adam Smith (1976), may be important extra-terrestrially in bringing together an understanding of those people affected by terrorism and those that are not and those that live terrestrially and those that don't to target terrorism and to prevent it spreading from its roots (Skinner and Wilson 1975).

It is likely that several countermeasure strategies will be implemented into policy to combat acts of terrorism in space. It will only be by the effectiveness of their application over time that it will be shown whether they are successful. But because of the nature of terrorism, those involved will develop methods to ensure that they are always one step ahead of the authorities. The breakdown of the economic structure of a settlement community leading to an extra-terrestrial economic crisis and a breakdown of capitalism could lead to an upsurge of terrorist attacks of various types as a means of seeking justice from the effects of the economic crisis (Cain 2015).

7.8 Alien Terrorism

One area of concern is terrorist attacks from aliens or other species that may have quickly evolved from humans due to the evolutionary selective pressures being applied extra-terrestrially or from artificial intelligence to gain supremacy in one or more space sectors (Lele 2003). Such "terrorists" could violently attack Earth-based settlements to prevent them spreading into their territory or use other means to prevent the spread of Earth cultures and civilisation. It is also possible that there maybe a sinister side to "alien terrorism" because of their superior technological development. They may use their superiority to blackmail extra-terrestrial settlements and Earth governments so as not to take resources from their territory in specific space sectors and possibly areas of the Moon, Mars and the moons of the outer solar system. A new paradigm will need to be developed to manage this situation for it may mean that Earth based civilisations and associated culture will be prevented from "colonising" space and be restricted to allocated interstellar space sectors (Molten 1978). But ironically, such restrictions on the development of Earth and its settlements will be to generate terrorism against the alien societies that could result in centuries of war; unless intelligence and co-operation amongst all species was used to prevent such a catastrophe.

7.9 Conclusion

Acts of terrorism have been used by all societies throughout history to further their causes and to instil fear into the targeted populations. This is likely to continue extra-terrestrially as Man explores space and establishes settlements with their own or federal governing bodies to try and ensure freedom and liberty for all. The death tolls that could be inflicted in space by an individual terrorist or by terrorist groups in pursuit of their causes are likely to be high because of the sophisticated nature of the weaponry that could be used and the devastation to settlements should their life support systems be targeted. The means to counter terrorism both on Earth and in space such as the use of surveillance and cyber security will need to be effective and this will entail some restrictions on the freedom of individuals to pursue their political, religious and scientific views. These restrictions, together with a strong enforcement body to deter terrorist acts and to implement harsh punishments for those conducting such acts, will need to be enshrined in a Human Rights Charter to apply both terrestrially on Earth and extra-terrestrially. The need for fair and just societies in space for all will be critical if terrorism is to be eradicated or more importantly never used or need to be used as Man conquers space.

References

Bruce, G. (2013). Definition of terrorism social and political effects. *Journal of Military and Veteran's Health, 21*, 26–30.

Burke, E. (1986). *Reflections on the revolution in France*. London: Penguin Classics.

Cain, J. R. (2014). Astronaut health-planetary exploration and the limitations on freedom. In C. S. Cockell (Ed.), *The meaning of liberty beyond Earth*. New York: Springer.

Cain, J. R. (2015). The Marxist theory of capital as applied to the economic development and governance of space colonies. In C. S. Cockell (Ed.), *Human governance beyond Earth. Implication for freedom*. New York: Springer.

Caruso, R. (2014). *Understanding terrorism: A socio-economic perspective*. Bingley, UK: Emerald Gp Publishing Limited.

Chaliand, G. (2007). *The history of terrorism: From antiquity to al Qaeda*. Berkeley: University of California Press.

Chatterjee, D. (2008). *Democracy in a global world: Human rights and political participants in the 21st century*. London: Rowman and Littlefield.

Chen, D. D., & MacCauley, M. K. (2011). In E. Sadeh (Ed.), *Commercial space actors in the politics of space: A survey* (pp. 104–119). London: Routledge.

Council of the European Union. (2005). The European Union Counter Terrorism Strategy. 14469/4/05, REV 4, November 30.

Crawford, I., & Baxter, S. (2014). The lethality of interplanetary warfare: a fundamental constraint on extra-terrestrial liberty. In C. S. Cockell (Ed.), *The meaning of liberty beyond Earth*. New York: Springer.

Crelinsten, R. D., & Schmid, A. P. (1993). *Responses to terrorism*. London: Frank Cass.

Dinerman, T. (2006, March 13). Hybrid wars and satellite vulnerabilities. *The Space Review*.

Dulo, D. A. (2014). Resilience engineering in critical long-term aerospace software systems: a new approach to spacecraft software safety. *JBIS, 67*, 150–158.

Gellner, E. (1983). *Nations and nationalism*. Oxford: Blackwell Press.

Goodin, R. E., & Jackson, F. (2007). Freedom from fear. *Philosophy and Public Affairs, 35*, 250.

Grenshaw, M. (1995). *Terrorism in context*. Pennsylvania: Pennsylvania State University Press.

Grothaus, N. (2014). *Types of Terrorism*. Hand of Reason RSS.

Hein, A. M., Pak, M., Buhler, C., & Reiss, P. (2012). Worldships-architectures and feasibility revisited. *JBIS, 65*, 119–133.

Held, D. (1995). *Democracy and the global order*. Redwood City: Stanford University Press.

Hertzfeld, H. R. (2007). Globalisation, commercial space and space power in the USA. *Space Policy, 23*, 210–220.

Hobsbawm, E. (2007). *Globalisation, democracy and terrorism*. London: Little Brown.

Hoffman, B. (2002). Rethinking terrorism and counterterrorism since 9/11. *Studies in Conflict and Terrorism, 25*, 311.

Hoffman, B. (2006). *Inside terrorism*. New York: Columbia University Press.

League of Nations Convention. (1937).

Lele, A. (2003). Space terrorism: Is it a possibility? *Institute of Peace and Conflict Studies*.

Listner, M. (2005, May 31). Its time to rethink international space law. *The Space Review*.

Molten, P. M. (1978). On the likelihood of a human intersteallar civilisation. *JBIS, 31*, 201–208.

Nagel, T. (1970). *The possibility of altruism*. Oxford: Clarenden Press.

Newlove, L. M., & Erickson, J. (2011). *Governance beyond the global: Who controls the extraterrestrial*. Montreal: International Studies Association Convention.

Orth, C. D. (1990). Interplanetary space transport using inertial fusion propulsion. In *UCRL-JC-129239, 9th International Conference on Emerging Nuclear Systems*, Tel-Aviv, Israel, June 28–June 2, 1998.

Pedahzw, A., & Ranstop, M. (2001). A tertiary model for countering terrorism in liberal democracies: the case of Israel. *Terrorism and Political Violence, 13*(2), 1–26.

Peoples, C. L. (2008). Assuming the inevitable? Overcoming the inevitability of outer space weaponization and conflict. *Contemporary Security Policy, 29*, 502–520.
Pogge, T. W. (Ed.). (2001). *Global justice*. Oxford: Blackwell.
Rathgeber, W. (2008). *The European architecture for space and security* (ESPI Report 13). Vienna: European Space Policy Institute, October 1.
Rawls, J. (1993). *Political liberalism*. New York: Columbia University Press.
Remus, N. L. (2009). The need to counter space terrorism—a European perspective. *ESPI Perspectives, 17*, 1–10.
Schmid, A. P. (1994). *Political terrorism: A research guide*. New Brunswick, NJ: Transaction Books.
Schmid, A. P. (Ed.). (2011). *Handbook of Terrorism Research*. London: Routledge.
Schmid, A. P., & Jongman, A. J. (1988). *Political terrorism: A new guide to actors, authors, concepts, databases, theories and literature*. New Brunswick, NJ: Transaction Books.
Schriener, T. M., & El-Genk, M. S. (2013). Inherently safe and long-life fission power system for lunar outposts. *JBIS, 66*, 388–394.
Sen, A. (2010). *The idea of justice*. UK: Penguin Books.
Skinner, A. S., & Wilson, T. (Eds.). (1975). *Essays on Adam Smith*. Oxford: Clarenden Press.
Smith, A. (1976). *The theory of moral sentiments*. Oxford: Carenden Press.
Thomson, A. (2006). Satellite vulnerability: A post cold war issue. *Space Policy*, 19–30.
United Nations. (1948). *Universal Declaration of Human Rights*.
United Nations Crime Branch. (1992). *Definition of terrorism*. http://www.unodc.org/unodc/terrorism_definitions.html
Wells, H. G. (2005). *A Modern Utopia*. New York: Penguin Classics.

Chapter 8
Architects of the Revolution: Design Philosophies to Maximise Extraterrestrial Liberty

Stephen Baxter

Abstract Extraterrestrial habitats impose limits on human freedom through confinement, isolation, and a dependence for survival on advanced communal technology. Is it possible purposefully to design such habitats, or elements of them, to maximise freedom, even to enable 'safe' revolutions against corrupt central authorities? But to plan an artificial environment to ensure 'liberty' for others is probably itself an oppressive act, in that some freedom of choice of those others must be removed. In the long term a solar system filled with diverse communities of all kinds may provide refuge for the oppressed. In the short term the most effective technological enablers of personal liberty may be a suite of key inventions, possibly simple ones, that allow the possibility of individual mobility and escape. This is never a perfect solution; small isolated communities may themselves be unhealthy. However, if one is not contained, one cannot be oppressed.

Keywords Space colonisation · Space habitats · Worldships · Governance · Extraterrestrial liberty · Science fiction

8.1 Introduction

> [Inguill] had always been able to grasp the key importance of maintaining control, in the empire of the Sapa Inca. Especially in a habitat like this, huge yet finite and fragile, where you had to control the people in order to ensure the maintenance of the complex, interlocked systems that kept them all alive. .. Of course this great box of a habitat – a box from which there was no possibility of escape, under constant and total surveillance from Hanan Cuzco at the hub, from the Condor craft that continually patrolled the axis, and from operatives dispersed on the ground – lent itself to such control. Baxter (2014, 285–6)

S. Baxter (✉)
British Interplanetary Society, 27-29 South Lambeth Road, London SW8 1SZ, UK
e-mail: sbaxter100@aol.com

C.S. Cockell (ed.), *Dissent, Revolution and Liberty Beyond Earth*,
Space and Society, DOI 10.1007/978-3-319-29349-3_8

The premise of this series of books is that extraterrestrial habitats will present challenges to human freedom through confinement, isolation, and a dependence for survival on essential communal technology. Since humans will expire in seconds if deprived of oxygen, the control of the supply of air in an enclosed, isolated environment is central to the question of extraterrestrial liberty.

Perhaps the most familiar dramatisation of how life support systems can be used as instruments of oppression is the Arnold Schwarzenegger movie *Total Recall* (1990, dir. P. Verhoeven). A 'Mars Federal Colony' is ruled by Administrator Vilos Cohaagen, whose only priority is the continued supply of minerals from the mines of Mars, and whose use of the air supply ensures total control: 'If you want to breathe you have to buy his air.' When revolution foments in a red-light district called Venusville, Cohaagen has no hesitation in cutting the air supply to the district altogether, accompanied by memorable images of duct fans spinning to a stop, and, metaphorically, gasping goldfish stranded on a carpet. In the end Schwarzenegger's memory-addled hero Douglas Quaid saves the day by starting up alien atmosphere-generating machines: 'If Mars had an atmosphere he'd [Cohaagen] lose control.' A rapid terraforming follows, so that freedom is achieved by *entirely removing* the dependence of the colonists on technologically supplied air.

But this is a pleasing fantasy. In reality (see Sect. 8.4) even the largest habitat—even a terraformed Mars—would not free its inhabitants from dependence on technologically maintained life support. And even short of such drastic scenarios, extraterrestrial habitats will impose constraints on freedom in subtle and surprising ways. Grant (1984), in the context of worldship designs, studied the stability of populations in such habitats under various controlling regimes. Since in terms of the long term stability of an isolated habitat, falls of population, causing a loss of capability and genetic variability, are as damaging as resource-depleting population excesses, a conscious linking of birth and death rates seems mandated. Grant speaks of 'birth forcing' as well as birth 'licences'; it may be necessary essentially to force people to have children.

In such environments of necessary confinement and control the danger of tyranny is an ever-present. But, following the example of Douglas Quaid, is it possible to use technological solutions to mitigate the challenge to freedom of an extraterrestrial habitat?

As opposed to sociological speculation on the motives behind quests for liberty, the emphasis in this paper is on possible engineering choices that might serve to protect freedom.

8.2 Rebellion in a Space Habitat

In the course of Earth's history, one means by which freedom has been sought from tyrannical regimes has been through violent revolution. Could such rebellions be waged in extraterrestrial habitats?

8.3 Weaponising the Technosphere

A tyrant in control of an extraterrestrial habitat could readily control a population using threats to withhold the air supply. Conversely for rebels to attack such central systems—in Fogg's term (1995, 327) the 'technosphere' of systems that maintains the habitat's viability—could be disastrous.

Robert Heinlein's *The Moon is a Harsh Mistress* (1966), about the rebellion of Luna City, a penal colony on the moon in 2076, dramatises these issues. The immediate challenge faced by Heinlein's rebels is their utter dependence on communal life support systems. The prison's Warden controls these essential systems centrally, from his physically isolated and heavily guarded complex. The rebels do understand that it is suicidal to strike against the infrastructure itself: 'The woman had been in The Rock almost all her life … yet could think of something as new-choomish as wrecking engineering controls' (44). In the end Heinlein's colonists commandeer the colony's central control systems by making an ally of a sentient computer, and disable the Warden's Complex by subverting *its* oxygen supply.

While Heinlein's colonists refrained from a self-destructive attack on their colony's essential systems, they—and the Warden before them—did weaponise them. Even very large habitats will be dependent upon a 'technosphere', for they will require conscious management to stay viable. Would it be possible to wage a 'polite rebellion', in which neither the rulers nor the rebels engage in tactics that might threaten the viability of the habitat itself? Or will such potentially lethal tactics as weaponising or even smashing the life support always be too much of a temptation?

The lessons of history are not encouraging. Even though there are rules of war which have proven effective, such as encoded in the Geneva Convention, throughout history humans have shown a willingness to use the supporting environment as a weapon, even at the cost of wrecking it. For example, one can attempt to deprive one's opponents of life support. The Harrying (or Harrowing) of the North was a series of campaigns waged by William the Conqueror in the winter of 1069–1070, in which famine was deliberately used to subjugate northern England: 'The King stopped at nothing to hunt his enemies. He cut down many people and destroyed homes and land … He ordered that crops and herds, tools and food be burned to ashes. More than 100,000 people perished of starvation. I have often praised William in this book, but I can say nothing good about this brutal slaughter. God will punish him' [Orderic Vitalis, 12th century; quoted in Bartlett (2000)]. Crops and herds, tools and food: William destroyed one year's harvest, and the means to raise next year's harvest. This was a genocidal act.

Conversely one might trash one's *own* life support lest the opponents use it. A 'scorched earth' policy is a military strategy which involves destroying anything that might be useful to the enemy while withdrawing from an area, especially food sources. It was used for example by Joseph Stalin against the German Army in the Second World War (Carell 1964).

On Earth, in the course of the wars of the past, one could usually expect the damage caused by such attacks to be limited at least on a planetary scale, and, with time, for the biosphere to regenerate. Neither of these assurances are available when waging war or rebelling within an artificial life support system: the destruction of the air supply, whether deliberate or as collateral damage, could kill both tyrants and tyrannised in seconds.

8.4 Arming the Technosphere

If humans could not be trusted not to use essential systems either as instruments of oppression or targets in war, could we elevate the technosphere out of human control altogether? One could imagine removing essential life support from the realm of threat either by modifying the habitat's passengers, or its mechanisms.

Perhaps deliberate harm to the technosphere could be made the subject of deep taboos, instilled from birth in a colony's inhabitants. Possibly this could be backed up by hypnotic, pharmaceutical or other conditionings. The problem with this approach is that anyone who somehow evades the conditioning—even a visitor to the habitat not previously subject to the conditioning—is in a position to rule.

Alternatively, could a sufficiently smart technosphere, governed by some advanced AI, be given the power to defend itself against any human manipulation or attack? This would essentially mean handing some policing powers to the smart infrastructure.

In Varley's *Steel Beach* (1992), set in the 23rd century, mankind has been expelled from Earth by aliens known as the 'Invaders', but survives on the moon and other worlds of the solar system. Varley depicts a rich lunar civilisation suffused by high technology. People live in what they describe as a 'free society' with a government that is 'small and weak', but only because 'many of the instrumentalities of repression have been gradually given over to machines—to the Central Computer—not without initial trepidation, and not without elaborate safeguards' (101). In the book this goes calamitously wrong when the AI itself breaks down.

Reynolds' *Blue Remembered Earth* (2012) dramatises similar ideas. In a 22nd-century 'Surveilled World', humans operate in a tech-soaked environment—their own bodies are dense with implants—and their behaviour is monitored by a 'Mechanism' which intervenes if antisocial acts of various kinds are intended. Usually a harmful act is simply stopped from taking place through the implants, following which the would-be perpetrator is put through a process of analysis, treatment and rehabilitation. But more extreme sanctions are permitted: 'The Mechanism would kill, if killing prevented the taking of an innocent life' (Chap. 22). In an extraterrestrial setting such a regime could well enforce the protection of essential life-supporting mechanisms.

The problems with these solutions are two-fold. A basic practical objection is that any such system of control will be inevitably imperfect; it may develop flaws, or it may be 'gamed' by some individual who could thereby use it as a means of oppression in itself. Reynolds describes small-scale, slow, subtle 'crimes' being committed outside the parameters of the Mechanism's programming. And there are fundamental philosophical problems in having one generation engineering either their own human successors or the environment they will live in so as to remove fundamental choices. The inhabitants have been put in a kind of prison in which they are allowed to run riot, so long as they don't wreck the furniture.

In a sense, to engineer for 'freedom' within a confined habitat is itself an oppressive act; the exercise is logically self-defeating.

But if one cannot rebel against the tyrant, could one at least escape?

8.5 Escaping the Tyranny: A Buffer to Freedom

As regards planetary colonisation, in a discussion of the habitability of Mars, Cockell (2002) notes that the oft-used analogy of space as being like the American West as a place to seek individual freedom is a poor one. On Earth the basic commodities that sustain human existence—oxygen, water and food—can readily be acquired over much of the land surface even by a single individual: there are always places to run to, and survive. As Cockell puts it, 'this provides an "independence buffer" against the excesses of the worst human beings society has to offer.' This buffer will not be available in an extraterrestrial habitat, within which every breath of air must be manufactured by communal systems, and outside which is only a lethal environment.

In his seminal work *The High Frontier* (1976), in which he sets out a progressive agenda for expansion into space, O'Neill similarly argues that mobility will be a buffer to freedom: 'In space, where free solar energy and optimum farming conditions will be available to every community, no matter how small, it will be possible for special-interest groups to "do their own thing" and build small worlds of their own, independent of the rest of the human population' (237). O'Neill is of course associated with large space-habitat designs such as the 'O'Neill cylinder' (64ff). O'Neill's work is however largely sociological in intent rather than technological; his space-habitat designs were stepping stones in a vision of a longer term drive into deep space, with small communities using the habitats as platforms from which to escape, and 'homestead' the asteroids in an analogy with the American frontier experience.

On a planetary surface and in space, what enabling technologies might be available to restore the 'independence buffer' of individual freedom of movement?

In the context of planetary habitats a self-sufficient mobile sub-habitat able simply to move away from a larger community—like a rover leaving a docking port—might be one means of escape. One obvious support for such mobility would be an imposition of standard interfaces for power, supplies and physical egress. If a rover or sub-habitat were able to leave one community and dock with another without technical challenges, its occupants' potential freedom of movement would be enhanced.

For such a sub-habitat to endure for long periods, an ability to exploit local resources for fuel and other essentials would be another enabler of independent mobility: for example a small, reliable, portable ISRU unit, able particularly to extract oxygen from the local resources. Roberts' *Jack Glass* (2012) is a bleak glimpse into a dystopian future of solar system colonisation, but a future supported by key enabling technologies, of which the most significant is 'ghunk', a foodstuff that will grow from hand-scattered spores in the presence of sunlight, water ice and asteroid rock (26)—a simple and cheap ISRU method.

A reliable, mobile power source would be another useful adjunct to independence. Depending on the environment simple solar-power technology is one possibility, perhaps with panels that could simply be unpacked or unrolled as desired. The movie *Back to the Future II* (1989, dir. Zemeckis) imagined the year 2015 as a strip-cartoon world of flying cars and hoverboards—but there is cheap and clean energy, from 'Mr Fusion' home energy plants, capable of processing household waste into fusion fuel. Small fusion plants may in fact be achievable in the near future. In October 2014 Lockheed Martin announced a high beta fusion reactor being developed at the Skunk Works, a cylinder 2 m by 2 m by 4 m—small enough to fit in a truck—may be at the prototype stage within five years (Shalal 2014).

Other obvious enablers of individual freedom could include cheap personal food synthesis machines, and cheap automated personal medical support. Perhaps the simplest and most effective freedom buffer of all (suggested by Cockell) would be an EVA suit simple and quick to don: more like modern Arctic cold-weather gear, perhaps, than space agency pressure suits, which require extensive prebreathing before use. To be able to 'throw on a coat' and simply walk out of a habitat on a whim would do a great deal psychologically to grant at least a sense of freedom.

In space, meanwhile, as noted above O'Neill imagined people having the freedom and technology to 'homestead' the asteroids (1976, 233), using large space habitats as a base, and outlined key enabling technologies for this to happen. Essentially, a home-made spacecraft built in space, without the need to climb out of Earth's gravity well, is a much more plausible project than an Earthbound back-yard project [as in, for example, Heinlein's *Rocketship Galileo* (1947)].

A family based at a space habitat would be able to purchase or home-manufacture a 'space wagon': a spacecraft hull ~ 10 m in diameter, a rocket engine and fuel, and a closed life support system. O'Neill's wagon would at least capable of taking a family, as part of a 'wagon train', on a months-long journey to the target asteroid: 'about as long as your great-great-great-granddad took to get from Illinois to California' (242). The hull would weigh some 500 tonnes fully laden, including a shielding layer of 'sand' one foot thick and a central storm shelter. Interplanetary navigation systems would be required, as well as access to preliminary survey data to determine the asteroid's resources before choice of target. At the asteroid, habitats would be constructed from prefabricated panels, or constructed from local metals using sun-powered vapour deposition techniques: 'Within two or three years a sphere with a land area of more than a hundred acres for habitation … could be made in this way, mostly quite possibly by a housewife monitoring a control computer from her kitchen' (245).

The required mass of shielding material, incidentally, suggests that a significant advance in mobility might be the development of lightweight radiation shields for small spacecraft: perhaps easily shaped lightweight blankets of some advanced material, or some form of energy shield.

Of course the ability to escape from a tyrannous environment solves only part of the problem of freedom; a refugee needs somewhere to escape to.

8.6 Destinations: Room for Freedom

Historically Earth itself has provided 'room' to escape tyranny; the North American colonists were able to use the 'room' of their undeveloped continent to evade what they saw as the tyranny of the British, and begin their own experiment in civilisation-building. Yet Earth itself is finite; Earth itself is merely a very large and very well equipped 'habitat'. Is it possible that a large enough extraterrestrial habitat would provide room for liberty?

The background to modern space colony studies has been reviewed by some of the pioneers themselves, such as O'Neill (1976) (Chap. 1). Well-founded visions of human life in space dated back to the early twentieth-century works of Tsiolkovsky (1920) and Bernal (1929). However O'Neill's landmark study (1976) presented the first significant space colony designs, based on plausible materials and technologies, to build on the lessons of the first generation of space exploration, including the results of the Apollo missions to the moon.

As indicated above, in O'Neill's scheme, large near-Earth habitats would be merely the first stepping stones into space. Their initial economic justification would be to sustain large populations of workers who would build orbital solar power stations (SPS), the output of which would be sold back to the Earth. Arguing from a premise that 10,000 workers in space would be needed to kick-start a significant industrial presence there (116) O'Neill proposed as a model starter colony his 'Island One', a sphere ~500 m in diameter, rotating twice per minute to provide Earth-equivalent gravity. This would be constructed largely from lunar resources and would host 10,000 people living at urban population densities. Island Two would be an expanded version with a population scaled up to 140,000 people, and O'Neill's Island Three (64ff) was to be a pair of rotating cylinders each 32 km long and hosting a population of 20 million (69).

O'Neill did postulate larger structures as engineering possibilities, such as a fourfold expansion of Island Three, the largest available with 1970s materials (6), which could support some 160 million people. Habitats much larger are conceivable. The 'McKendree cylinder' was originally proposed at NASA's 'Turning Goals into Reality' conference in 2000 by NASA engineer Tom McKendree (2000). An expansion of O'Neill's Island Three built with such modern materials as carbon nanotubes rather than steel, the habitat would consist of two cylinders approximately 460 km in radius and 4600 km in length, containing 13 million km^2 of living space: nearly as much area as the nation of Russia.

Meanwhile, some insight into likely habitat types on planetary surfaces comes from a consideration of plans for the colonisation Mars.

A recent design of a near-future surface habitat on Mars was part of Project Boreas, by the British Interplanetary Society (Cockell 2005), a study of a base at the Martian North Pole designed to support 10 people through a Martian winter. As described in a paper on base design by Greene (p 32–48) the living spaces would consist of seven cylindrical mobile habitats. The usable habitable area would total 289 m^2, or some 29 m^2 per person, equivalent to some 35,000 people per km^2. This is equivalent to a high urban density; the base would feel seven times more crowded than New York. However scientific and industrial facilities commonly support temporary populations of significantly greater densities than the urban.

As regards more advanced Martian bases, Zubrin's *The Case for Mars* (1996) gave a detailed scenario for the progressive colonisation of Mars, with 'dozens or even hundreds of people within a few decades of the first landing' (218). With the first colonists living in 'tuna-can' landed habitats, Zubrin described a growing population housed in underground galleries of Martian brick, or in domes and spherical structures on the surface. A 50 m-diameter sphere with internal decks spaced 3 m apart would have a floor area of 21,000 m^2; 100 people living in such a sphere would live at a population density of an urban-like 4800 per km^2. Clarke's *Sands of Mars* (1951) depicts a still more mature Martian colony, Port Lowell in the Aurorae Sinus, consisting of six 'large domes' (83), each 250 m in diameter. With 2000 people (88) and an inhabitable area of at least 0.3 km^2 (there is some habitation underground) the colonists live at a reasonable urban density of 6700 persons per km^2. The colony is austere: 'With its rows of uniform metal houses and few public buildings it was more of a military camp than a city, though the inhabitants had done their best to brighten it up with terrestrial flowers' (91).

On planetary surfaces, very large settlements are conceivable. Taylor's 'worldhouses' (Taylor 1992) would be constructed by lifting a transparent, airtight roof 1–3 km above the Martian surface by means of towers spaced some 6 km apart; large areas of the surface could be enclosed incrementally, and a breathable atmosphere gathered within. The argument is that scarce volatiles could be concentrated in a limited space rather than spread across the global surface. If a worldhouse were erected over Hellas Basin, for example, the 2300 km-wide crater would afford a living area of some four million square kilometres—covering about 3 % of Mars's surface area.

While O'Neill argues about the 'evils of bigness' (1976, 39) including political oppression and high crime rates, there are strong technical and social arguments for building large habitats. Fogg (1995, 483ff) argues that it may be impossible to scale down Earth's biosphere by many orders of magnitude (five orders from Earth to O'Neill's Island Three) and expect it to maintain all its functions adequately. Davies (2013) has recently outlined more subtle problems that might arise from an attempt to construct a limited biosphere. On Earth humans are part of an intricate ecological web which is dominated, in biomass, by microbes. We don't know how this web could be 'truncated' to fit into a smaller environment than Earth's: which species would it be 'safe' to leave out?

Meanwhile, in contrast to O'Neill's opposition, social benefits of large habitats have also been suggested. Asimov, in his novel *Nemesis* (1989), suggested that perhaps humans need a certain population size to develop and sustain civilisations: 'No civilisation ever developed except in a continental area, or in islands in proximity to a continental area …' (292). And Robinson points out (2012, 332) that revolutions seem to require spaces to gather: '[Historically] physical masses seem to matter. If enough of the population takes to the streets in mass demonstrations, governments have no good defence'.

However all of these closed environments will require technological support and conscious intervention. This may apply even on a fully terraformed Mars. As Fogg notes (1995, 327), as Mars lacks such features as tectonic cycling, for very long term stability 'terrestrial-type geochemical cycles may have to be closed with the assistance of technology. Unlike the biosphere of Earth, that on Mars may require the presence of a civilisation for long-term survival. Technosphere and biosphere must coexist in symbiosis.' (This incidentally contradicts the resolution of *Total Recall*, which showed a dictator's control broken when Mars is rapidly terraformed.)

Thus even the largest extraterrestrial habitat will suffer from the sociological flaw explored in the course of these books: a dependence of its inhabitants on centralised systems, and thus a natural tendency to tyranny. The great Inca habitat Yupanquisuyu of the author's novel *Ultima* (2014) is a McKendree cylinder, so large that its atmosphere is a comparatively thin blanket coating the inner hull. Its rulers dwell in axial shelters protected by the vacuum, its hull is saturated with surveillance and other technologies, and the inner habitable surface, all of which is continually visible, is patrolled by aircraft. Yupanquisuyu is a habitat the size of a continent, but in the past comparatively technologically primitive regimes, such as the Mongols, have controlled empires on such scales.

However it seems likely that there will not be one space habitat in the future, however large, but many.

8.7 The Freedom Spectrum: A Colonised Solar System

> [Gnaeus] produced images of structures in space, grainily realised, cylinders and spheres and wheels, more angular structures. "They cluster around the major planets, or trail them in their orbits around the sun. And they come in all sizes, from units the size of small Roman towns, Centurion, to much larger. There may of course be smaller constructions below our ability to resolve. .. And although this 'Quechua' is their dominant language, evidently the official one, we hear scraps of many other tongues – including bits of Latin."
> Baxter (2014, 268–9)

An early vision of a fully colonised solar system was given by Bernal in *The World, the Flesh and the Devil* (1929), a remarkably prescient survey of ultimate human possibilities: 'On earth, even if we should use all the solar energy which we received, we should still be wasting all but one two-billionth of the energy that the sun gives out. Consequently, when we have learnt to live on this solar energy and also to emancipate ourselves from the earth's surface, the possibilities of the spread of humanity will be multiplied accordingly …' (17). Out there, we will have the energy and resources to spread out.

What might life be like in a heavily populated solar system? Day to day, with booming populations in confined artificial environments, one would expect a life of continual hard work just to ensure survival. Ashworth (2012a), in papers on the developmental background to a worldship programme, noted (149) that 'space colonies are to be understood, not a small artificial self-regulating planets, but as large self-contained buildings requiring continual maintenance. An astro-civilisation must be constantly active in order to survive: repairing space colonies, scrapping superannuated ones, recycling their materials and constructing new ones, just as an urban civilisation does on Earth today.' Indeed Ashworth argued that urban population densities would be more appropriate to habitats requiring such support.

It seems probable that ultimately far more humans in a populous spacefaring civilisation will inhabit space colonies rather than live on planetary surfaces. Cockell (2002) has noted that Mars, while offering the most Earthlike conditions available away from Earth in the solar system, is (quoting Captain Scott) still an 'awful place to live', suffering sub-Antarctic temperatures, lethal radiation regimes and a thin, unbreathable air, among other hazards and discomforts. Arguing from Arctic and Antarctic analogies Cockell estimates a total population on Mars of some 0.5–4 million people—unless the planet is terraformed, when the population could rise to $\sim$100 million.

Meanwhile we should not expect a widely colonised solar system to be politically stable, or static. Robinson's *2312* (2012) contains a recent fictional depiction of a future of a 'Swarm' of space habitats. At this point in time the solar system has a population profile still dominated by Earth, with 11 billion inhabitants (307). By comparison, of the colonised worlds Mercury has half a million (310) inhabitants and a partially terraformed Venus 40 million (510). A terraformed Mars is also populous. The space habitats, while numerous, have a total population of $\sim$500 million. Politically Robinson's solar system is fractured. Earth and Mars are significant powers; Venus, en route to terraforming, is dominated by the Chinese; the inhabited asteroids meanwhile are united in 'an economic system of nested co-ops organised for mutual support' (125). But the situation is seen as continually turbulent because of a shifting of power, potential and population. There is a growing divergence between Earth and the spacegoing (73); a fully terraformed Venus will be a much more significant player.

With time, fundamental limits of physics and humanity may lead to the emergence of distinct groupings. In Clarke's elegant *Imperial Earth* (1975), set in 2276, Earth and its colonies are divided by lightspeed delays. Communication delays are trivial on Earth, leaving that world relatively unified; but the long delays in speaking to the colonised worlds reduce effective interpersonal contact and so reduce human unity (101). The story concerns a young citizen of Titan making a once-in-a-lifetime trip to Earth to make the face-to-face contacts that still, according to Clarke, are a necessary underpinning of human society. If Clarke is right, one could imagine the finitude of lightspeed, coupled with an ancient aspect of our humanity, fragmenting us into blocs, each highly connected internally: Earth-moon, Mars, the Jupiter and Saturn moon systems, perhaps some clusters of asteroids.

What are the ultimate limits to this process? How many people might live in a fully developed solar system? O'Neill (1976, 247) speculates that the resources of the asteroids, equivalent to the surface area of 3000 Earths, might fill up at modest growth rates in four thousand years. O'Neill's 3000 Earth equivalents (1976, 247) might support 10^{13} people. Savage (1992, 271), Lewis (1996, 196) and Ashworth (2012a, 149) give convergent estimates in the range 10^{15}–10^{16} people. It is clear that the vast majority will be living in space habitats, not on planetary surfaces. As to when these maximal limits might be reached, recalling O'Neill's prediction of a few thousand years, Ashworth (2012b) (158) notes that given a starting base of the consumption of 1000 t of asteroid material per year, and a growth rate of 1.4 %—lower than historic human population growth rates, but sufficient to give a doubling time of 50 years—the asteroids would be consumed after some 2500 years.

Even if we do not race to such limits so quickly, these studies indicate that there is indeed room in the solar system for an expansion of human possibilities almost beyond the imagination. In a heavily populated system one would expect habitats of a range of sizes, from family homesteads to populous 'cities', underpinned by political and ideological frameworks of all kinds, perhaps gathered in 'nations' isolated not by geography as on Earth but by lightspeed delays.

Somewhere out there, one would think, there would be a haven for the most desperate of refugees. This may be the ultimate technological solution to the problem of extraterrestrial tyranny. But whether we reach this future is in our own hands.

8.8 Conclusions

Extraterrestrial habitats, with their populations isolated and heavily dependent on artificial life support systems, are naturally tyrannous environments. The subject of this paper has been on possible engineering choices that might serve to buffer freedom and independence. The tentative conclusion is that until the sky is filled with habitats as destinations for escape, the best assurance of freedom may be enabling technologies, possibly quite simple ones, which allow individuals or groups to escape from environments under repressive control.

For a utopian visionary Gerard K. O'Neill is remarkably pessimistic about humanity. 'Mankind does not change,' O'Neill says (1976, 18). 'At the most we can suggest opportunities whose technical imperatives will make it easier for mankind to choose peace rather than war; diversity rather than repression'. And one can see his vision of endless 'homesteading' expansion and escape, all the way to the limits of the solar system, as a working-out of flaws in our human nature: an inability to live peacefully in a large society, an inability to control our numbers or our consumption of the environment's resources.

Another gloomy utopian has suggested, however, that a future in space may always be too difficult for us. In Robinson's *2312* (2012) the spacegoing have their own doubts about the fitness of their society for its environment: 'Living in space may be too hard for us. .. When you combine political inadequacy with the physical problems of being in space … we may be trying to make an impossible adaptation out here' (230).

There are however alternatives to a future of growth and escape. A *steady-state society*—in which economic activity is contained within levels which the local biosphere is capable of assimilating and regenerating—is economically and physically at least a feasible solution to the conundrum of survival in any closed environment such as a space colony. O'Neill (1976, 27) himself points out that steady-state societies are possible, and cites the pre-Conquest Inca empire as an example. However such a society, says O'Neill, must be 'rigidly structured, dictatorial'; 'almost any static society is forced in self-defence to suppress new ideas … at his death [a peasant in the Inca empire] … left a world almost exactly the same as the one he was born into.' Zebrowski (1979, 225) expressed similar sentiments: 'Dogma goes hand in hand with a lack of growth, both in knowledge and in technical ability, while growth *must always* accompany democratic ideals of improvement and innovation' (my italics). But these are American writers, working in the tradition of the Western frontier.

In the modern world, the question of how to build a prosperous global society *without* economic growth is receiving considerable attention (Jackson 2009). Some of this concern was prompted by the financial crisis of 2008, but concerns about the limits to growth go back to John Stuart Mill, and were starkly raised by the Club of Rome report in the 1970s (Forrester 1971). The transition to a new society of 'stasis' would require new macro-economic models with different relationships between firms and consumers, new types of governance emphasising the pursuit of long-term social and ecological goals—and, most significantly, a shift in societal values towards components of prosperity other than the acquisition of more and newer goods, such as meaningful participation in social life. Jackson quotes experimental but small communities searching for such answers.

Possibly by the time we inhabit large colonies in space we may necessarily have had to learn how to live within steady-state societies on Earth without compromising our essential human freedoms; with new goals and new motivations, such ways of living might be readily transferred to colonies in space.

If however we cannot learn to get along with each other—without a fundamental advance in human nature—it seems that space habitats may always be naturally tyrannous environments, and too fragile to contain rebellions. In that case it may be that the only way technologists can help to ensure freedom will be by providing key enabling technologies to allow means of escape, both from closed, tyrannous habitats, and from our own flaws.

References

Ashworth, S. (2012a). The emergence of the worldship (I): The shift from planet-based to space-based civilisation. *Journal of the British Interplanetary Society, 65*, 140–154.

Ashworth, S. (2012b). The emergence of the worldship (II): A development scenario. *Journal of the British Interplanetary Society, 65*, 155–175.

Asimov, I. (1989). *Nemesis*. New York: Doubleday, 1989. Page numbers from the 1990 Bantam edition.

Bartlett, R. (2000). In J. M. Roberts (Ed.), *England under the norman and angevin kings 1075–1225*. London: OUP.

Baxter, S. (2014). *Ultima*. London: Gollancz.

Bernal, J. D. (1929). *The world, the flesh and the devil*. London: Jonathan Cape. Page numbers from the 2010 Prism Key Press edition.

Carell, P. (1964). *Hitler moves east, 1941–1943*. New York: Little, Brown.

Clarke, A. C. (1951). *The sands of mars*. London: Sidgwick and Jackson. Page numbers from the 1986 paperback edition.

Cockell, C. (2002). Mars is an awful place to live. *Interdisciplinary Science Reviews, 27*, 1193–1198.

Cockell, C. (Ed.). (2005). *Project boreas: A station for the martian geographic north pole*. London: British Interplanetary Society.

Davies, P. (2013). Afterword. In J. Benford & G. Benford (eds.), *Starship century*. New York: Microwave Sciences.

Fogg, M. (1995). *Terraforming*. New York: Society of Automotive Engineers Inc.

Forrester, J. W. (1971). *World dynamics*. Cambridge, Mass: Wright-Allen Press.

Grant, T. J. (1984). The population stability of isolated worldships and worldship fleets. *Journal of the British Interplanetary Society, 37*, 267–284.

Heinlein, R. A. (1947). *Rocketship Galileo*. New York: Scribners.

Heinlein, R. A. (1966). *The moon is a harsh mistress*. New York: Putnam's. Page numbers from the 1969 NEL paperback edition.

Jackson, T. (2009). *Prosperity without growth? The transition to a sustainable economy*. Report for the sustainable development commission. Summary http://research-repository.st-andrews.ac.uk/bitstream/10023/2165/1/sdc-2009-pwg-summary.pdf. Accessed Nov 18, 2014.

Lewis, J. S. (1996). *Mining the Sky*. New York: Addison-Wesley.

McKendree, T. (2000). Implications of molecular nanotechnology technical performance parameters on previously defined space system architectures. In *Turning goals into reality*, NASA, 2000. http://www.zyvex.com/nanotech/nano4/mckendreePaper.html#RTFToC17. Accessed Nov 18, 2014.

O'Neill, G. K. (1976). *The high frontier*. New York: William Morrow. Page numbers from the 1978 Corgi version of the second edition.

Reynolds, A. (2012). *Blue remembered earth*. London: Gollancz.

Roberts, A. (2012). *Jack glass*. London: Gollancz.

Robinson, K. S. (2012). *2312*. London: Orbit.

Savage, M. T. (1992). *The millennial project*. Denver: Empyrean Pub. Page numbers from the 1994 Little, Brown edition.

Shalal, A. (2014). Lockheed says makes breakthrough on fusion energy project. Reuters. http://news.yahoo.com/lockheed-says-makes-breakthrough-fusion-energy-project-123840986–finance.html. Accessed Nov 8, 2014.

Taylor, R. L. S. (1992). Paraterraforming: The worldhouse concept. *Journal of the British Interplanetary Society, 45*, 341–352.

Tsiolkovsky, K. (1920). *Vne Zemli* (*Beyond the Planet Earth*), translated by K. Syers (1960). New York: Pergamon Press.
Varley, J. (1992). *Steel beach*. New York: Putnam's. Page numbers from the 1993 HarperCollins edition.
Zebrowski, G. (1979). *Macrolife*. New York: Harper & Row.
Zubrin, R. (1996). *The case for mars*. New York: Simon & Schuster.

Chapter 9
Collective and Artificial Intelligence and Their Role in Dissent and Revolution Beyond Earth

Lewis Pinault

Abstract Engineering for Liberty—spacesuit, habitat, and social communications designs as espoused by Charles Cockell for helping ensure better safety, fairness and opportunities for creative dissent—is examined in light of current New Space endeavours and their Space Agency and regulatory context. Its approaches are contrasted with recent science-fiction based speculations and put in the context of fast-emerging Collective Intelligence and Artificial Intelligence capabilities.

Keywords Airbus · Artificial intelligence · Asteroid mining · Astrobotic · Collective intelligence · Engineering for liberty · Extraterrestrial liberty · Human governance beyond Earth · International space authority · Lunar mission one · Lunar resources · Mars exploration · Moon · New space · OneWeb · Space commercialisation · Space exploration · SpaceX · World space agency

9.1 Engineering for Liberty: A Better World Than Aurora?

In the last volume, I expanded on ideas supporting an International Space Authority (ISA), which as I re-emphasise here is very much meant as an enabler of distributed, collective, and diverse exploration activities, rich in the fullest possibilities of liberty and dissent beyond Earth. Here, I pick up editor Charles Cockell's own themes on planning for liberty in space—in fact an important role for an ISA to play.

In his chapter 'Extraterrestrial Liberty: Can It Be Planned?', Dr. Cockell offers practical steps toward advancing a degree of robust liberty beyond totalitarian control, combining insights to human needs, drives and tendencies with potential mitigations for the extraterrestrial environment's biological and resource constraints. Dr. Cockell's chapter merits full exploration in its own right, but for purposes of examining how an 'Authority' and Collective Intelligence systems can

---

L. Pinault (✉)
UCL/Birkbeck Centre for Planetary Sciences, London, UK
e-mail: l.pinault@ucl.ac.uk

C.S. Cockell (ed.), *Dissent, Revolution and Liberty Beyond Earth*,
Space and Society, DOI 10.1007/978-3-319-29349-3_9

contribute to the kind of dissent that makes for safer and better societies in space, I paraphrase his key proposals here, and begin by drawing attention to a key new science fiction work that offers perhaps important counterexamples.

[Spoiler Alert: This Chapter gives away key suspenseful elements of Kim Stanley Robinson's finely written *Aurora*—as *The Guardian's* Adam Roberts puts it 'the best generation starship novel ever.' Readers are encouraged to enjoy the full adventure first!]

The potential need for totalitarian decision-making is starkly developed and drawn to dire conclusion in Kim Stanley Robinson's 2015 novel *Aurora*. It is a remarkable turn for the author who gifted humanity with a wondrous, better-than-Earth saga of human expansion in *Red*, *Green*, and *Blue Mars*, culminating in the fully inhabited Sol system of *2132*, where the main threat to human existence appears to be the development of Artificial Intelligence.

In *Aurora* we are treated to the despairing view that the Earth ecosystem's many interdependencies are un-replicable, that long-duration voyages by generation starships will inevitably lead to harmfully differing rates of evolution by cohabiting humans, bacteria, and other species, and that other planetary systems, rather than lending themselves to terraforming, seem likely to introduce life forms inevitably inimical to humans, dramatically in this work in the form of a fast-working mad-cow-disease-like prion.

For the purposes of these volumes with Dr. Cockell exploring human liberty and governance, it is instructive to look at the responses of the starship crew at each of the major crisis points of its journeys. Early in the voyage to their 1000-year distant target world of Aurora, twin co-journeying starships, between whom periodic biological and social exchanges were meant to mitigate risks for both crews, meet catastrophe. One of the giant ships, complete with a dozen biomes and 1000's of humans for whom starfaring lacks both choice and the potential enjoyment of either origin or destination planet, is completely and instantly destroyed by the act of single terrorist, apparently by mistake.

The remaining starship crew, understanding what happened to its twin from its final data transmissions, chooses to close ranks around its elder council, and supported by its increasingly self-aware AI (who refers to itself as 'Ship'), elects to 'erase' the recollection of a second starship and their own implicit vulnerability to terrorism by never referring to the event, never teaching the episode to future generations, and completely excising all mention of it from public records, as well as controlling and filtering all media feeds from the fast-receding and increasingly time-delayed Sol System so that the facts of the twin's existence and demise remain with a very select few, each generation.

Dr. Cockell offers a provocative series of proposals for 'engineering' in favour of liberty in the extremes of the extraterrestrial environment, proposals that perhaps might help avoid the mishaps of the Aurora-bound crews. He notes how important freedom of movement is to humans generally, and how with a ready supply of comfortable spacesuits and interchangeable parts and supplies to support them, spacesuits can become enablers of freer movement and personal exploration and choice. So too with space habitats: by providing a distribution of habitats engineered for ease of ready access, both safe harbour and redundancy to ensure against catastrophe ease the sense of total dependence that centralised habitats and cumbersome, more controlled-access spacesuits would engender. In the fearful world of *Aurora*, all movements are highly constrained and pervasively monitored.

Dr. Cockell develops the role that an independent Extraterrestrial Oxygen Board can play in ensuring open competition in the supply of this key resource, and goes on to suggest how incorporating as much as possible of the means of oxygen production into a given machine helps ensure both safety and fairness. He raises how planetary surfaces like Mars where oxygen may be directly extracted from the atmosphere and other sources also mitigate in favour of distributed access, suggesting that these might be priority developments over centralised oxygen distribution and control.

Where his work on engineering for liberty redounds to both Earth and space environments is raising the dangers and means to avoid an all-pervasive surveillance state. By encouraging a culture of open information, and developing newspapers and their equivalents as a diversity of trusted and accountable voices, he proffers a series of safety checks that can both help improve the odds against catastrophic disaster, and quality and liberty of the societies evolving in extraterrestrial environments.

9.2 Are the World's Space Agency's Already on the Path to Liberty?

Dr. Cockell's proposals for engineering for liberty suggest a different, more collectively driven and sustained set of premises for space exploration itself. Rather than relying on a single central national agency like NASA that in the main continues to control the full gamut of exploration objectives, mission selection, means of transport and habitation, detailed design, and selection and training of crews as well as a veritable army of carefully vetted and dependently funded researchers and institutions, with the likes of Space Exploration Technologies (SpaceX) now a more emergent approach can be better envisaged, and the benefits of managing a rich and richly distributed array of concepts and approaches more fully appreciated.

By winning the right from NASA to meet specific goals for transport to the International Space Station (ISS) through its own design schemes, Elon Musk's SpaceX has opened to the door to the emergence of new patterns and structures from the complexity of all possible approaches to reaching beyond Earth's gravity well. He has made re-usability a respectable goal once more for wider, lower-cost access to space, regaining credibility for the idea of one-day routine journeys for which the Space Shuttle so painfully raised hopes—and then summarily lost. It may not be his first-stage rocket controlled vertical descents that drive the needed order-of-magnitude cost savings; it could be fellow billionaire Jeff Bezos and Blue Origin.

By creating the atmosphere where re-usability seems a practicable premise, a famously profit-driven company like British Aerospace can see its way to a major investment in Reaction Engines, paving the way to realise the British Interplanetary Society dream of a Skylon space plane. With its breakthrough advances in heat exchange, Reaction Engines has perfected the means to draw oxygen from the atmosphere, rather than carry its weight to ignite fuel. This means that a Skylon-like space plane need not be in a hurry to get to the upper atmosphere, and therefore can take advantage of lift as well as thrust. Once beyond the oxygen-rich heights of the atmosphere, the engines only then convert to full rocket-powered thrust, meaning that the whole ensemble of winged spacecraft and dual-capability engine is much lighter weight overall—enabling another path to full re-usability.

So too now in the satellite business. Even at speed of light transmission, increasing data manipulation at 40,000 km geostationary orbit (GEO) heights appreciably lags behind applications that are served from low-Earth orbits (LEO) of mere 1000's of kilometres. But to provide the same coverage from much lower altitudes, a fleet of 100's of satellites is required rather than a handful; and because they fly in the thicker realms of the Earth's atmosphere, they will inevitably see their orbits rapidly decay—which means to keep an operating full-coverage fleet or 'constellation' of LEO satellites going, the whole lot has to launched quickly, which in turn means they must be mass-produced to be at all cost effective.

9.3 New Space and Regulatory Opportunities for Liberty and Dissent

These are not new ideas—reusable spacecraft and satellite constellations are concepts at least as old as the space age itself. But in the hands of NASA, ESA and their largely military-needs driven institutional contractors, the need and impetus to bring them to fruition has been absent, and the idea of widening the idea and player base seen an unnecessary complication at best. In a remarkable sign of the once unimaginable partnerships to come, the Airbus Group's Defence and Space division has teamed with the Google-nurtured entrepreneur Greg Wyler and his LEO satellite fleet company OneWeb, with funding from Virgin Galactic's Richard Branson. With rapid design and prototyping to be done at Airbus headquarters in Toulouse France, drawing on aircraft assembly line expertise, the mass-manufacturing enterprise will be set up in the United States. Thus, driven by both consumer and military demand for internet and bandwidth (largely for drones), a 'New Space' business is being born.

The present-day major obstacles to unleashing a vast array of new partnerships and alliances is the lack of appropriate and responsive regulatory frameworks. A spaceplane returning to Earth for a runway landing, because it has elements of both thrust and lift, is subject to two different sets of regulation; in the UK, where Richard Branson is hoping to bring his Virgin Galactic spaceplanes to land in the

UK's own spaceports, this resolution of overall governance has been slow in the making. Similarly, the major hurdle for OneWeb at the moment is not design-to-cost, mass assembly, or extensive new supply chain—Airbus has all these remarkably to hand—but instead the nation-by-nation negotiations required to win the global coverage the OneWeb fleet is designed to deliver. In the previous volume *Human Governance Beyond Earth* I offered a rationale for an International Space Authority based on the International Seabed Authority convened by the United Nations and now successfully awarding mining rights under the high seas; in my chapter 'Toward a World Space Agency' I even suggest this could be the basis for a semi-independent body to best accommodate the integration of private commercial interests with scientific and environmental concerns and requirements.

It may be that we are already seeing signs of a more open framework for space exploration and development that will lend itself to higher degrees of liberty and safeguard dissent. As Europe's national space agency's become more active and closely partner with commercial interests of all scales, the supra-body of the European Space Agency is becoming more a Research and Development enabler and focal point and partnering entity for Russian and potential Chinese collaborations. Similarly in the USA, with SpaceX leading the way on taking on more of the detailed design tasks, NASA may too evolve into more of a research enabler. With the likes of a semi-independent International Space Authority or similar to provide a framework for national and commercial partnerships and their extraterrestrial endeavours, a baseline for a more distributed approach to human liberty and governance beyond Earth is perhaps already foreseeable.

9.4 Collective and Artificial Intelligence in the Service of Extraterrestrial Liberty and Dissent

There remain two major factors, to my view, that may go the longest way to enabling and benefitting from a diversity of ideas and dissenting opinions and ways of living beyond our home planet: Collective Intelligence, and perhaps the more counter-intuitive, Artificial Intelligence.

I mean Collective Intelligence to embrace the idea that an ably selected group of people will almost always make a better decision or judgment than even the most intelligent or experienced individual in that group. As a one-time moderator for MIT's 'We are Smarter than Me' initiative and having worked with crowdsourcing initiatives from innocentive.com to openIDEO I can assert that there are devils in the details of how a group is selected, how challenges are set, and how collaborations are facilitated on line—but nonetheless I observe that the growing weight of the social science literature and findings demonstrate that we make better decisions together.

At Lunar Missions Labs, we're in the business of proving this out, reaching to our KickStarter backers. They successfully got us going on the path to landing an automated craft on the Moon to carry a time capsule of humanity's stories (and perhaps much of the planet's DNA), and now we are turning to this base of 1000's of largely non-space professionals for challenges ranging from landing site selection to drilling options (for future capsule burial) to how we might communicate the archive's contents to future discovers, be they our descendants or others. This is an experiment whose outcomes are yet to be seen, but the energy and interest would be the envy of any space agency's marketing and public outreach teams. Lunar Missions recently announced a partnership with Google Lunar XPrize leading contender Astrobotic, and the prospect of carrying a first trove of personal digital stories to the Moon in less than two years' time is only outweighed by the excitement over prospects that we may be able to turn over control of the landers six high-definition cameras to 'the crowd' on the platform during select periods of the lander's operations.

Collective Intelligence thus raises the prospect that we may all become space explorers, in person and through our automata—not as passengers and watchers but as idea-driving active participants. Just as Mars One or some similar reality-TV driven enterprise may radically broaden the idea of what being an astronaut and settler of other worlds may mean, we may have only begun to tap opportunities for a more massively participative path to extraterrestrial society. NASA's Jet Propulsion Laboratory's behind-the-scenes support for realistic visualisation for the box office hit *The Martian* has already spawned a Fox Virtual Reality simulation for would-be extraterrestrials, and both the sensation and reality of settling other worlds may soon be more of a world, rather than agency, nation, or industrial concern.

It may seem early to speculate on Artificial Intelligence's impacts on extraterrestrial dissent, but some of the relevant stakeholders, notably the Oxford Institute for the Future of Humanity's Nick Bostrom and SpaceX's Elon Musk, are already giving the prospects—and dangers and potential benefits—of AI's rapid evolution deep attention. Just as Dr. Cockell has proposed that we may engineer for liberty and dissent in our extraterrestrial lives, Prof. Bostrom offers models for setting Initial Motivations with 'superintelligences' that we are already in the process of inventing. He notes that while humans have no 'top-model' operating principles, it is reasonable to expect that the very manner of their creation may make our superintelligences more receptive to setting initial motivations, specifically for philanthropic friendliness and the motivations to stay friendly. This is both more than might be said for humans, Bostrom notes, and a potentially humanity-ending omission if we do not act quickly to instill these motivations.[1]

[1]In my own research, I am developing computer vision and machine-learning capabilities to assist in the Search for Extraterrestrial Intelligence (SETI) by way of seeking evidence of trace artefacts, including in our own solar system. There is active debate within our own UK SETI Research Network about whether we should broadcast messages 'to' ETI's rather than just seek signals from them; again since our own future 'superintelligences' may be the stewards of communications either way, one might hope that they—and 'them' might have Bostrom's philanthropically friendly top goals.

Just as some of the world's most intensive research and investment in AI, autonomous systems, and data analytics is today led by the aerospace industry, future superintelligence applications may first be brought to bear to help manage the immense complexities and autonomous requirements of the human settlement of extraterrestrial environments. If the 'Ship' of Kim Stanley Robinson's *Aurora* is to prove as friendly to humans as it is portrayed to be (and as helpful as it *must* be in all the dire circumstances Robinson creates for his ill-fated travellers), then there is work that likely needs to start sooner than later. We can be rightly proud of what we are beginning to achieve through human Collective Intelligence, and by using this as a demonstration of who we are, we can begin the modelling for a superintelligence of what our superintelligence inventions should be friendly 'to', in all our glorious mix of drives and desires. As Bostrom notes, 'in humans, with our complicated evolved mental ecology of state-dependent competing drives, desires, plans, and ideals, there is often no obvious way to identify what our top goal is; we might not even have one.'

The risk is that we engineer for a superintelligent top-model that is not friendly—e.g. for the pervasive surveillance that Dr. Cockell aims to help us avoid as extraterrestrial settlers. And unfortunately, we may not have a lot of time to address the question, before say a giant asteroid collision brings a prompt end to all human extraterrestrial ambitions. As planetary impacts specialist Dr. Adrian Jones noted in a recent UCL lecture, the fact that we witnessed the Earth-annihilating scale of the Shoemaker-Levy comet impacts on Jupiter in our own lifetimes when as a species we've only *just* started to be able to see such things should have us on very high alert indeed. It is perhaps the irony that the best asteroid avoidance systems that can be invented with sufficient urgency and timeliness will themselves be AI-driven—will they be equipped for top-model friendliness, or perverted to weapons creation? It is up to our Collective Intelligence to make the right choices and challenge and engineer the ability to offer dissent for the wrong ones.

Chapter 10
Art and Change in an Extraterrestrial Society

Annalea Beattie

Abstract In places where human rights will be constantly be under threat, for instance in vulnerable, off-Earth communities living in extreme environments in the future, will art have the potential to support us to break out or withstand? How will it enable us to share alternatives that challenge or disrupt those who subdue us? Drawing upon a fleeting, firsthand experience of art making, one that leads from the field of astronomy beyond art to a moment of political change, art in this context is situated as highly adaptive and evolving, as one node of political reality that shapes and is shaped by its social and political conditions. As I experiment to take the contingent qualities of a poetic experience of drawing in the dark into prose, this paper explores how art connects to the politics that surround it—how art has the capacity to change us.

Keywords Activism · Art · Immersion · Change · Cultural resistance

10.1 Preface

Close up there are fingermarks in the rise and fall of paint—expressive smears as well as deft brushstrokes, all of which take some boldness and some planning. A step back and the self-portrait takes up all the frame, facing out, chin up, eyes locking eyes. What grips me here? This painting is by a condemned man I have never met. At the time a war of words broke out. Signed by one hundred and fifty thousand people, a petition for mercy read, 'if you don't accept anyone can change, there is no incentive to change.' (Mercy Campaign 2015). Add the thoughts of Alphonso Lingis (1998, 175), 'It is tomorrow that gives its sense to whatever I do today; it is tomorrow that gives its sense to whatever I am today'. But there is no clemency now.

A. Beattie (✉)
RMIT University, 124 Latrobe St., Melbourne, VIC 3000, Australia
e-mail: annalea.beattie@rmit.edu.au

C.S. Cockell (ed.), *Dissent, Revolution and Liberty Beyond Earth*,
Space and Society, DOI 10.1007/978-3-319-29349-3_10

Australian drug trafficker Myuran Sukamaran became an artist on death row. While imprisoned for a decade in a Balinese gaol, he turned to art as a way of comprehending and challenging the conditions of his incarcerated world, painting portraits of his captors, memories of his family, studies of other inmates and of himself. As his art practice developed, Myuran was mentored and in turn he mentored others, convincing prison authorities to set up art classes for prisoners as a form of rehabilitation and raising money for a gallery within the prison from the sales of his own work. These art classes were the first gatherings in the prison's history where men and women could interact, transforming the prison community.

10.2 Political Activism and Art

Does art transform us? I believe it does. But before we can consider the potential of art in an extraterrestrial society in the future, what can be learnt from the relations between political activism and art here on Earth?

When we think about how revolution and art associate, there is a common tendency to recall easily read propaganda from the past, evoking as Viennese art theorist Gerald Raunig (2007, 69) suggests, a 'recurring conceptual misinterpretations that activist art repeats the totalizing urge of some political art of the early twentieth century'. Even so, we are not likely to return to the dynamic red and black, to sheaves of wheat, to the upraised fist or peasant youth and beauty as monumental symbols of power and dominance. These are not our contemporary images of radical openness or of interdependence. Just recently in Tehran, the famous Islamic Republic propaganda billboards of 2013 were removed. Described as 'so aggressively decorative, radically inclined towards surface that they became almost self-referential', the billboards offered, 'an unlikely blend of political portraiture, religious kitsch, Russian Constructivism and heavy metal.' (Zolghad in Bradley and Esche 2007, 446). Full of stale visual rhetoric and empty of any political content, the billboards have since been replaced with posters advertising community-based art competitions. As global capitalism creates an endless mushrooming of powerful representations, the point is not everyone believes everything they see or all of what they read. And this will most likely be the same on Mars. In terms of visual representations the revolution is over.

Yet there is an ongoing debt that political activism owes art production and this is worth remembering. In its endless forms, art has been implemented as a tool for state or official persuasion, whether it perpetuates or opposes violent or prolonged, cruel, unjust treatment. Significantly, it was the legacy of propaganda art that evolved to inform a century of art activism, transformed by artists within movements such as the Situationist Internationale, Conceptual Art and Feminist Art. It reached its zenith in the conservative Reagan era as American artist organisations such as ACT UP, Gran Fury, Group Material and others, used posters, slogans, billboards and advertising to highlight the body in political art, struggling against the lack of resources and the terrible silence that surrounded racism, discrimination and the invisible diseases of AIDS. Art forms evolve and good art can run fast, outlasting its social context. My favourite modernist artwork is Bertelli's speedy *Continuous Profile—Head of Mussolini* (1933) and then there are all those Old Mistresses, still going strong after centuries.

In our developed world, if we are removed from concepts that constitute the word 'revolution', we are more familiar with the hegemony of forms that already exist in the field of human rights. This thinking can be extended to how art might responsively reframe demands that inhere in the discourses of not only human but non-human rights. As we shift position to think about art and future communities in poisonous surroundings far from Earth, the focus is still one of social responsibility. The key question is, if art is seen to be one kind of political reality, how does it embed itself both in politics and in the production of knowledge and power?

To explore how an experience of art making can activate a frame of reference that addresses this responsibility through our potential for change, this paper now turns to an expression of subjectivity in flow—an encounter with the physical and the non-human through drawing in the dark.

10.3 The Dark Skies Project

The **Dark Skies Project** (2015) focuses upon urban light pollution in the night sky and requires a kind of willing blindness. In this instance, its context is a space science field trip called New Zealand Spaceward Bound, part of NASA's international outreach programme and the first project of the University of Auckland's new Institute of Astrobiology.

10.3.1 Context

In Aotearoa, in the Taupo volcanic Zone, a region that has features analogous to planets like Mars, New Zealand, Australian and NASA scientists use scientific research to mentor other scientists, as well as teachers and space science students, questioning how life evolved over billions of years and what life is like or what it might be like on other planets. There were fifty of us from all over the world welcomed into the Te Takinga Maori community with song. Each of us told our mountain and our water, now we all sleep together on woven straw mats in the hand-painted red, white and black marae. Directly above, we gaze up into the stellar nursery where young stars are born, straight up into the twinkling bright and jet black barred spiral galaxy that is the Milky Way, with its massive one hundred billion planets and thousand billion stars. During the day our group explores the limits of microbial adaptation in the geothermal springs amongst the volcanic landscapes of the McKenzie Basin. At night, when the moon sets, I am here to draw past the bazillion stars to the dark sky.

10.3.2 The Bortle Dark Sky Scale

In the darkness outside the marae, I invite our space science group to draw the amount of light in the sky. If we hold our arms high perhaps we can use the Maori compass—our hands and our fingers, to section off a part of the night sky. We look past the stars to observe a measure of light in the hue of the sky with our eyes, as opposed to seeing it through the astronomer's lens. Though we are in the dark, we rely upon observation, drawing what sky glow we see, focusing on tone and intensity of light more than the stars the sky contains. Inside the artificial light shows us what we have drawn. As a scientific method of interpreting the level of light in the night sky with the naked eye, we can use *the Bortle Dark Sky Scale* (Bortle, J. 2006) to name a grade in the amount of light in our drawings. The Bortle Scale helps us measure the degree of light pollution in the sky and compare the keenness of our visual acuity.

10.3.3 Why Is Darkness Important?

In our Australian cities as unwanted light emits above the horizon, it scatters off dust or water particles into our atmosphere, impeding vision and inhibiting our view of the night. Light pollution makes the entire sky light grey or orange (Bortle 2006). When we stand in our urban back gardens or in the city street at home, the sky is often too bright to see the Milky Way and there are few visible stars at night. For instance, in Melbourne, of the five stars that make up our national icon the kite-shaped Southern Cross, sometimes the sky is so full of urban light we can only count four.

Darkness is precious to many living things. When obtrusive light pollutes the dark, it extends the length of the day. The trees adjust their seasonal variation, the foraging behaviour of marsupials changes, as does the mating and gestation patterns of birds and insects. While we contemplate the lovely dappled lights reflecting on the water, silver gulls circle above the bridge in loose flocks, feeding all day and all night. In the bushes by the creek, bright light spills over from the freeway, inhibiting the call of the male tree frog and he ceases to call out for his mate. The green sea turtle hatchling is disorientated as she navigates towards the glare of the security lights shining on her natal beach rather than towards the starlight shining on the waves breaking on the sand. Here in Australia, possums are dazzled by urban light. The irises of their eyes become permanently damaged by floodlights shining from the street up into their tree houses.

And too much light disturbs our ecology. Every year the Australian bogong moths rely upon the earth's rhythms of night and day as they fly low from the black soil plains of Queensland to their traditional breeding grounds in the Snowy Mountains. Travelling only at night, bogong dislike light. When their path crosses through the massive beams of upward lighting placed around the base of Canberra's

Parliament House, they instantly drop to the ground, searching for a resting place to escape the brightness. Thousands of moths descend on Parliament, squeezing under doors and through air conditioning ducts, settling on ceilings and windows frames. As they crawl into the dust under cupboards, they press their nutty bodies into every crevice, hiding, seeking to bury themselves, their soft bark-brown forms littering the carpets.

10.3.4 In Darkness

What does the darkness mean for us? Here we are together in the dark. Four hundred thousand kilometres away, the cold dry orb of the moon moves together with us, locked in an embrace with our tidal pull, always showing us the same lovely, lucent face. Each night we return to what we know of it. This night, fast-running clouds race to cover the second full moon of the month—blue moon. In the few short moments before it is shielded from view, we enjoy the play and nuance in the rays of rainbow light streaming from its hazy hue. Dark comes. Colour leaves the landscape, altering it to grey-green-blue-grey. The remaining light trails across the ground, hiding and showing contours, our bodies softened, enveloped in shadow. The fingers of the darkest shadows move over us, closing in and obscuring the bush. Vision dims, our forms dissolve, our gender, our egos drop away.

In the blackness now the stars shine loud. Eyes strain and are stopped by dimness as we absorb the loss of light. Cool breezes sweep by and we draw our clothes together. On this now moonless night we stand mute against the southern sky and look out across the horizon. Hear crack, rustle, call. The wildlife scatters but the night sky continues silence. Though we can no longer see our feet, through the soles of our boots we contact firm ground, our faces turn up, we crane our necks, adjusting and orientating our stance. We stretch tall and like the gums, feel the vertical strength in our spine.

Our childhood points to where the Iron Pot is located on the celestial equator. Only a thousand light years away we say and tell the others how the constellation was created, arbitrary, unrelated, consisting of red and white-blue super giants much larger than our sun, collapsing in on themselves, burning out and lonely slowly dying. Or perhaps we pause and lift our arms, extending to place one palm on the Southern Cross and one on Achernar—roll off the tongue *Al Ahir Al Nahr*— and as we bring our hands together to meet in front of our bodies, there is Tasmania, there is the deep south of Antarctica, there exactly is the South Pole.

But this is enough. We tire of ourselves and go back to darkness. So dark. Our sentient gaze is drawn upwards, invoked by distant elements and endless, boundless depths. We follow faint auroras, nebulae, the steadier lights of planets, a sudden flash and a long comet tail to the north. To the south, sparkling Sirius calls us. Searching for movement in dusty galaxies, we wander to lose ourselves with long distance travellers in the deep. Wondering, immersed, our gaze softly lights and drifts. Wrapped in the dark, in the unseen and inaudible beyond our experience, we become as trillions. Motionless, staring, we see our reflections sensually grasping and groping at the stars and hear the ephemeral echoes of our listening wafting back on the currents of air. Then in the same breath, our disengaged image returns to earth, saturated to the skin. A moment ahead of ourselves in a point in time not yet begun, we alone see ourselves looking at the night sky, just as we alone feel the damp clothes clinging to our legs.

Yet in the wet bush amongst the smells and the shapes and the continuum of sound, we are not solitary or friendless. In the weak zodiacal light, the Milky Way casts diffuse shadows on the ground. As we step forward, the stars drop a fraction closer to the horizon, stars drop down. We throw back our heads and roar. The dolorous soughing of the wind that lifts the leaves and the tendrils of hair speaks back soothingly to us, the sound of squeaking, creaking, scraping limbs of trees, the periodic whhoooooo—*wit* of the fledging olive whips, the faint far terror of the devil screaming in the distant scrub. Something dead in the nearby brackish water reminds us of something. At our feet beneath the moss, the high, rhythmic pulses of the cricket interrupt and grreccc grrreccc the frog sings out. The leaf litter of the silver eucalypt pats to the ground one leaf at a time.

10.3.5 Drawing Darkness

In this balmy New Zealand star wonderland, does urban light really disrupt the black sky, upsetting our bodily rhythms and that of the animals, birds and insects? Perhaps it might be Venus, Mars and Jupiter in a straight line just above the western horizon but I'm never sure, it's very late. I hold my paper high and begin to draw, following closely the chromatic tones of the nocturnal sky. Unpredictable on all insides the night closes in. Drawing traces of light, I look past ensembles of stars to where the shifting dark light confuses, turning into distance, into doubt, into the future. Closing one eye, I advance with the other, squinting into the darkness, into the *Monocular Abyss* (Snow, M. 1982), hearing the rush of my exhaled breath and see it, feel it, puff haaaaaa into the air. I close both my eyes. The tone of the sky floats in my vision like the thin sum of all my encounters with every night, strikes me hard then falls away, sky memory fading as I lean towards it. When I look back the night sky is cold, stranger beyond all I could ever know, closed within the extremities of its extraterrestrial self, outside my grasp.

Distracted by refracted light more than its lack, the graphite works in negative, drawing black. It scribbles, rasps and scrubs at the rag, scratching as it traces trajectories. The pencil anticipates and chases the light, intuiting depth by imagining it, drawing it out and throwing it down, each moment different. My eyes keep pace with the flow of changing light, roll back and forth from sky-paper-sky. Right eyed, I avert my gaze, shifting my eyes to the side as astronomers do, and placing the sky portion I have chosen on the nasal side of my vision, attempting to find peak sensitivity. In the equilibrium of my looking, the dim paper is indistinct, colour the wave-length of grey hair though I know its surface is changed as I am changed by the night light. I forget intent. I try not to hold my breath as I approach the limits of my vision, pressing my eyes against the sky. I press my thumbs against my eyes, cosmic flashes flare, fade and ***.

As the surface of the paper begins to darken, I notice in my throat an absorbing trance-like pulse, a beat, a pace, a flow of blood, as back and forth my upper arm repeats its narrow angle. At the same time eclipsing vision is an insistent, small, immediate sound—the meeting of pencil impacting on paper and paper resisting. The zig-zag-zig-zag rise and fall scratch scratch scratch cht CH cht CH cht CH is prolonged, confident and faster in tempo when building up a ground of dark carbon on the surface of the paper, then slower, steadier and more hesitant when drawing light or is it drawing none? Temporally expanding, this movement of the pencil lies between the earth and the dark sky. This sound of intense contact augments and then it becomes the full-bodied activity of drawing in the dark.

Does it matter if I can't see the drawing in the dark? As I draw my earnest gaze is drawn. A soft mist comes in, blending my sight with the sky. Contented, the night participates to draw itself as abundant carbon in the stars, comets, in the atmospheres of most planets, within the pencil and within me. I blacken the character of the paper and let my vision go. I am sober, there is nothing else—only what my acute body makes of the drawing.

10.3.6 Thinking Through Darkness

In the amorphous dark light, does drawing fix meaning or does the darkness conform to my rules? I am not, as Alphonse Lingis says, 'grasping the carpentry of things in their appearances nor subjectively fabricating images of them', (1998, 101). Nor am I occupied by their intricacies, their resemblances or their concepts. Instead my intentions are at once distracted by and keenly conscious of where images are leading, only images, 'their shadows, their reflections, halos, the harmonics of their colours, the rhythm of their forms' (1998, 101). Unmoored, there is nothing else to cling to. No art is needed to mimic the spectacular resonances of the common space that is the sky. No face to face encounter is necessary, no ocular-centric supremacy of the eye, no revelation or sublime moment of insight as phantasmagoric spectacle, no wordy, deterministic framework that catches and grapples meaning and drags it down to safe ground, nor is there left behind any bias or idea of finitude, not even the visible and invisible. Though we can trap the immensity of the night, stake it out on a curve and net its coordinates, even if we can somehow snatch at its radiance through the data provided to us by our senses, there is no representation in words or images that tells us how it all ends. Our destiny in the night sky is banal and carnal, as anonymous and indeterminate as the past and future of every animal.

Absolute, the dark sky is the roof of our home. In it, we greet the alien companion always with us, always within us. Without void or gap or pattern or witness, with no schema, no terrifying exteriority, no words, no meaning, we take the sky into our core. We inhabit it as we are part of its space. We intuit the field that extends beyond us, around us, coming in towards us. The sky is beneath us. If we turn away it is still there. Behind my lids it is still there. In it is our ordinary day, our past and future.

There is how. Return to the dialogue between the drawing and the darkening night sky. Just as I find the tone of the sky, the drawing disappears in my hand. Perfect, it is there but not there. The drawing has become the sky. The sky becomes it and they are each other. And there is not a thing to show, only the failure to really comprehend the event, a coming to nothing that takes me beyond perception into the realm of my experiences of the night, as drawing that immerses me in the darkness so that I continue it.

10.4 Art, Responsibility and Change

So how might these experiences of drawing the night sky translate? Just now as I write I can hear the soft flapping of moths bashing against the glass lamps outside my door and sense the lesser stars in the starry national park above my home. Believe me, I am the murderer of moths. I am responsible. As a maker of light pollution who creates more light than is needed within my own urban environment, each night I stop the dark—the back light on. Inside, I artificially light my world for the after-dark freedom and the snugness of seeing our faces and our homes in soft never-ending glow. I affirm safety and security with everything in its place, a fortress with every space lit, no strangeness in my house, the clocks visibly ticking out their emphatic moments until dawn. Turning night into day, I am the one who chooses to interrupt and curfew the night. But I can alter my own nocturnal environment and advocate for an ancient sky brimming with stars. I am changed by drawing the black night and switch the lights off.

By observing the night sky and drawing its gradient of light, we attend to vital relationships between seeing and knowing, seeing and surveying, as well as the curtailing of sight and the role of light in shaping visual experience. This affective encounter offers potential for resistance as we give importance to dark skies and to the experience of art with no art in sight. As we acknowledge the role light pollution plays in our visibility of the stars and in our ecosystems, in our energy consumption and in our health, there is the opportunity to act differently, mobilising daily forms of resistance and self-determining organisation in response to light pollution. Drawing, when is used to ideate, is an empowering vehicle for us to translate movement and experience into feeling and action, predicated on the act of making. Drawing the light in the sky or its lack can expand thinking and lead to a kind of delayed understanding about ourselves, agitating, as Anja Kanngieser suggests, systems of value (2012, 284). Unconnected to one community or to individuals, this event provides an uneven, temporary meeting point that gives form to a provisional act, one that privileges our different, ordinary experiences of the black night and provokes a sense of belonging. In drawing the night we belong to the universe and

Fig. 10.1 Annalea Beattie (2015) 38.1378°S, 176.2514°E, dark skies project (24 × 14 cm, pencil on rag)

to *each other*. This idea is as old as we are, built on community. As anarchist Landauer (2005, 165) said, 'The state is a condition, a certain relationship between human beings; we destroy it by contracting other relationships, by behaving differently towards one another' (1910). This is not revolution but it is cultural resistance. Every time we see the dark sky, we image our own footprint and we connect the heavens with our planet to situate the self as part of the vigorous web of forces that make up our world. In switching the lights out we advocate for the dark night, for ourselves and all our ecosystems within it (Fig. 10.1).

When we leave the security of home and the warmth of our fires to travel to other planets as explorers or even as refugees on starships, we will be constantly challenged by our new circumstances. We may be menaced by those who control our oxygen and possibly that we could end up living as spectators to our own imprisonment, trying to exist amongst people who are aggressive or broken. On Mars, as we contend with the threat of real danger, pain or harm, anxiety and the continual tension in shared space could lead to deteriorating morale, to stagnation or depression, to a loss of wellbeing in management terms and for us, damage to human health.

Pointing beyond discourses of violence, on Mars, resistance through art will experiment with models of organisation that follow change and resist fixity. To disrupt the all powerful representations of the state, formal and aesthetic criteria may be no longer be enough as a foundation for describing what revolutionary art is and what it can do. This means we may be compelled to behave in ways that position our relations in the world as material and partial, 'in which transitions, overlaps and concatenations of art and revolution become possible for a limited time, without synthesis and identification.' (Raunig 2007, 18). To avoid subjugation, self-interest and powerlessness in our community far from Earth, art will have the capacity to challenge existing paradigms as it traverses territories that both humanise and pay attention to the limits of our human selves, building solidarity and resistance in daily experiences, inciting change.

10.5 Postface

Through exhibitions organised by fellow artists, his friends, his family and through social media, Myuran Sukumaran's prolific art practice in an Indonesian gaol drew attention to his situation and that of the other prisoners, stimulating public debate, prompting mass vigils and outrage. Finally the Australian Government and the European Union were persuaded to jointly petition Indonesia to declare a moratorium on capital punishment. As the violence of his execution drew closer, Myuran continued to working on the last of his art works—a large bleeding heart and a portrait of himself surrounded by darkness. His final request was that he be able to paint until his death. Myuran Sukumaran claimed art transformed him and gave him the will to live. How this moment of change can arise uninvited from art making has been the guiding thought for the writing of this paper.

References

Bertelli, R. G. (1933). *Continuous profile (Head of Mussolini)*. courtesy of London: Imperial War Museum.

Bortle, J. (2006). *The bortle dark sky scale*. http://www.skyandtelescope.com/astronomy-resources/light-pollution-and-astronomy-the-bortle-dark-sky-scale/. Accessed September 8, 2015.

Bradley, W., & Esche, C. (2007). *Art and social change—A critical reader*. New York: Hachette Book Group.

Kanngieser, A. (2012). The transversal politics of performative encounters. *Deleuze Studies, 6*(2), 265–290. doi:10.3366/dls.2012.0062.

Landauer, G. (1910/2005). Weak statesmen, weaker people, der sozialist. In R. Graham (Eds.), *Excerpted in anarchism: A documentary history of libertarian ideas—Volume one: From anarchy to anarchism (300CE-1939)*. Montreal: Black Rose Books.

Lingis, A. (1998). *The imperative*. Bloomington: Indiana University Press.

Mercy Campaign. (2015). http://www.smh.com.au/federal-politics/political-news/150000-australians-sign-mercy-petition-for-chan-and-sukumaran-20150215-13f9j6.html. Accessed May 10, 2015.

Raunig, G. (2007). *Art and revolution transversal activism in the long twentieth century*. Cambridge: MIT Press.

Raunig, G. (2014). Singers, cynics, molecular mice: The politics of aesthetics of contemporary activism. *Theory, Culture and Society, 31*(7/8), 67–80.

Snow, M. (1982). *Monocular abyss*. Polyester resin, 111.8 × 69.2 × 139.7 cm. https://www.academia.edu/7091924/Michael_Snow_Objects_of_Vision_2012. Accessed October 28, 2015.

Chapter 11
Sturm Und Drang: Guaranteeing Teenage Liberty in Space—An Exploration of the Joys and Dangers of Adolescence in an Off World Context

Janet de Vigne

Abstract This chapter will address the emotional and relational issues concerning extra-terrestrial existence, focussing on the problems adolescence might pose to a contained community. It will consider the behaviour of children, possible origins of behavioural problems and possible solutions. It will consider too the neurobiological, psychological and biological development of the human and the effect this might have on behaviour; definitions of human maturity; ways of managing relations to the benefit of the community as a whole, and the consequences of all this for considerations of human identity and liberty. Adolescence will be considered in terms of collective responsibility and the emotional landscape of the group. Lastly, conflict resolution, reconciliation and relationality will be explored with regard to children's and young people's liberty and well-being in an off world context. Sturm und Drang is used in the Stanley-Hall sense (Adolescence, Its Psychology and its Relations to Physiology, Anthropology, Sociology, Sex, Crime, Religion, and Education, Appleton, New York, 1904) to describe the anxiety, stress and pressure and striving of the adolescent attempting to negotiate the complex and searing landscape of emotion during the teenage years (Stanley-Hall in Adolescence, Its Psychology and its Relations to Physiology, Anthropology, Sociology, Sex, Crime, Religion, and Education, Appleton, New York, 1904). It is also of course a reference to the German literary period of the late 18th century, characterised by works of rousing emotion and rebellion.

Keywords Adolescence · Neurobiology · Education in space · Teenager behaviour · Off-world human development · Relationality and community in space · Educational liberty · Endocrinological liberty · Genetic liberty

Everyone reading this will have had experience of the teenage years—either personally, or as a parent. They are perhaps the most difficult period of human development to negotiate, and a time about which great stories are told. Comedians such as

J. de Vigne (✉)
Lancaster University, Bailrigg, Lancaster LA1 4YW, UK
e-mail: janet_devigne@yahoo.com

C.S. Cockell (ed.), *Dissent, Revolution and Liberty Beyond Earth*,
Space and Society, DOI 10.1007/978-3-319-29349-3_11

Harry Enfield have developed very successful teenage characters, whose behaviour on screen suspends rational belief in the way that only great comedy can. Kevin the teenager is rude, insolent, monosyllabic, unreasonable, aggressive, moody—passionate (when screaming at his parents) and/or sulkily silent (when in their company). (To see for yourself, find the numerous scenes still available on youtube). What a combination of raging hormones and confusion! His parents cannot cope—all their loving and rational responses are greeted with howls of derision and aggression. Kevin 'works' as a comedy character because although these traits are taken to extremes, everyone recognises them. (This does not preclude of course the fact that some people's lived experience might be of something worse). How dangerous might a teenage crisis be in an off-world context? What can we now envisage happening from what we know, and what might we prepare for, without attracting charges of manipulation and a consequent loss of liberty? How much of the Sturm und Drang period is biological—and what is the nature/nurture balance? The discussion will of necessity cover the issue of emotion, as this is a major part of the landscape to be negotiated. In an extra-terrestrial setting such as this, the whole group must be considered—the effects of Sturm und Drang will ebb and flow; all emotions do.

In a contained context such as a space ship, or a potentially threatening environment such as a new planet with hitherto unknown hazards, moodiness and unreasonable behaviour become more concentrated irritants than in the communities of earth. The relationality of the group might be compromised, not just by its adolescents but by those who might not manage the inevitable challenges and testing of boundaries as the new seeks to overthrow or at least equalise the family structure. These issues are problematic—but there are other, more dangerous manifestations of the teenage that might pose problems fatal rather than awkward to the ship or the settlement. Biologically, the intra-developmental stage of a neurological restraining mechanism for risk is one; a propensity for addiction is another; but we should also consider the psychological need to establish an identity and find a place within the group. The management of an alpha male or female whose advent could not be predicted from her genetic make-up might cause upheaval until he or she is settled into a role everyone is happy with. On the Hebrides, remote islands off the west coast of Scotland, such an upheaval still occurs in the hearts and minds of the population. The Isle of Lewis is fairly remote and the population is not huge; surrounding nature is superb and people live happily with a three and half hour ferry journey to the mainland. However, at some point many begin to experience chronic high anxiety, viewed as a precursor to the 'curam'—a religious conversion to the Presbyterian patriarchal and rigid tradition that has hegemony. The choice is stark—to belong and obey, or to resist and go mad—as suggested somewhat controversially by Parman (1990). It is also expected, and people wait it out—almost like waiting for an addict to go 'cold turkey'. The human meta-narrative is peppered with such stories of adolescent crises of belonging. There are, of course, places to go on a beautiful Hebdridean island. There may not be anywhere to hide or commune with nature (as so many of us do) to find 'answers' in an off-world context. How then would it be possible to negotiate a similar situation in space? Could this developmental stage be 'headed off at the pass' while avoiding charges of manipulation? Could a space be found for a Vulcan-style kahs-wan maturity trial? (Not to be confused with the pon farr, although this is often a human issue too, and of course a fictional projection of our needs).

One of the many issues around teenage behaviour is their attitude to and use of the environment. Kevin has the freedom to slouch moodily in the park with his friends, sucking on an illegally obtained cigarette. He can go out to a shopping mall or cinema, sit on a car park wall, go to a skate park or mooch around the streets in his hoodie. We live with the irony of an ongoing controversy in our times—too much generally available space, but few, safe, designated spaces for teenagers. (Kevin would be very unlikely to join a youth group, or anything organised by adults). This is one of the factors contributing to a sense of alienation, the rise of gang culture, and extreme loneliness in our young people. Young men in 21st century still constitute a major risk group for suicide (the third leading cause of death among 14–19 years old, AAoP 2000); others, alienated to an extreme degree seem, ever more frequently, to resort to gun or drug crime as a way, perhaps, to punish society. Our young women too are at risk.

Risky behaviour among adolescents manifests in various forms. There is challenging behaviour, where authority is questioned; there is also competitive risk taking induced by peers. The latter is possibly the most dangerous—teen films are full of scenes of 'chicken' (a game played to see who might back off first as two cars speed towards each other). 'Daring' each other to do more and more ridiculous stunts is something that results in casualties every year. Often these incidents do not end well, and because of the digital age we are now in, there is a permanent web-based record of what has happened. This adds 'shame' into a mix that can quickly become powerfully toxic—it is today not possible to escape one's previous behaviour, forcing teens into a place of dread until some event might be exposed. Dread of humiliation, shame and a need to maintain one's position in one's peer group (a position hard won as teens seek to make and remake their identity) are very worrying factors to have in the wings as a small group of people play out human destiny off-world. Parents will need to exercise specific skill sets to help here, as will teachers—the understanding developing in the qualitative investigations of social science must be explored and lessons learned! The very idea of the 'son of Sheldon' in a crew on its way to Mars would give one the screaming heebie jeebies. Of course it is possible to argue that a Sheldon type would never be allowed onto a crew in the first place, but the potential for Sheldon-esque character development in a child cannot be blamed solely on genetics. Real geniuses may arise, but will need the social skills to function in the group, therefore their human behaviour (and that of the group, such as, say putting them on a pedestal or showing fear of their ability), must be carefully managed to avoid such narcissistic unpleasantness.

Everyone would agree that adolescence is difficult—how did Sheldon become the monster?—it's a time for testing boundaries, testing your strength, your relationships, intelligence, your understanding of the environment. There are complex neurobiological and endocrine-related reasons for this, of course. The period has been studied in depth and indeed the title of this chapter owes its name to the work of Stanley-Hall in 1904 (and before that to Goethe and Schiller, who were adolescents once…). Current research, with the added benefit of technology that allows us to see the brain without a dangerous intervention, is helping to explain some paradoxes and confirm some theories. The discussion here concerns whether we should consider adolescence as a possible threat in an off-world context, and prepare in advance some strategies for minimising this without compromising individual liberty. Contrary to popular belief (and that wonderful photograph of James Dean), it is by no means definite that rebellion and its consequent risky behaviour need inhabit the heart and mind of a teenager.

It is not so long ago that we demanded the ultimate sacrifice of our teenagers—the average age of young people dying in the First World War and subsequent conflagrations is an obscenity. As I write Europe is deep in remembrance of events in 1915, someone I got to know recently told tales of a great uncle who engaged the Red Baron in combat at the age of 19. An RAF pilot of that age had a maturity beyond his years forced on him. What kind of leadership skills might we need from teenagers in space? How can we judge maturity and trust a young person with responsibility for the group? So what, then, constitutes biological maturity, and how might this differ in terms of psychological and emotional maturity? How possible is it to engage with personal liberty—should we look at this as an individual or a collective issue? If we consider this in a Rawlsian sense, then if 'each person is to have an equal right to the most extensive basic liberty compatible with a similar liberty for others', people will have to work together to look out for each other's liberty in order to vouchsafe their own. Teaching the value of the team and understanding the relationality of human endeavour become essential if we are intelligently to manage and limit risk.

Many researchers now associate risky behaviour with emotive cues or stimuli, rather than exclusively developmental factors (Casey 2013a, b). The teenage brain is at a special stage of development. 'The social, psychical, sexual and intellectual challenges' of adolescence are 'biologically and experientially' moulding it (Casey 2013a, b). (Here it is worth recalling the work of Susan Greenfield on the hippocampus—as we learn, it enlarges, it enlarges, as we learn (Greenfield 2001). The process of brain development is therefore reflexive and dynamic—driven by experience (stimuli) as much as biology (the trigger of the organism to move into the new stage of life where reproduction is possible). A few years ago the thinking was different—the teenage brain is more developed at the back than the front, therefore it was thought that the self-regulating and control systems held in the frontal lobes were not functioning optimally (Casey 2013a, b)—the 'defective' teenage brain. The perfect excuse for irrational risk taking was born—your brain simply cannot 'do' the law of consequences. Where, then, in terms of biological determinism, can we place a theory of liberty that is appropriate? How can we work with the developing brain instead of against it? This is where choice and anticipated emotional outcome may be useful considerations.

Casey's review of current research shows that there are different types of trajectory in brain development (Casey 2013a, b). Where teenagers are concerned, there are peaks and troughs that are now considered as adolescence-specific, but there are also longer processes that begin in childhood and go through adulthood.

One of these is the cognitive control mechanism—this emerges in adolescence but keeps developing throughout adulthood. It is therefore not specific to adolescence per se, so to attempt to explain risky behaviour as solely something specific to a teenage lack of pre-frontal circuitry would be wrong. The more dangerous trajectory that runs alongside this is the brain's sensitivity to social stimuli or cues for more risk-prone behaviour that become attractive as the adolescent attempts to negotiate new social challenges. Steinberg's work [Gardner and Steinberg 2005 in Albert et al. 2013)] shows a correlation between peer interaction and risk among teenagers. In an experiment conducted among 14 year olds, 19 and 37 year olds involving a computerised version of 'chicken' (driving forwards and avoiding crashing into a wall that might suddenly appear) the teens showed a hugely increased propensity for risk taking when in the company of other adolescents —'it's no fun doing it on your own' (Albert et al. 2013). Risk at this age is then associated with the fun of partaking in a group activity. Joy seems to outweigh danger. Adolescents are not able to negotiate socio-emotional and choice related incentive cues' (Albert et al. 2013) and in fact, where the 'happiness' factor of the peer group is concerned, choices will not be made at the conscious level—'it seemed like a good idea at the time'. It seems therefore that over time, the cognitive development happening in the brain helps teens to desensitise themselves to peer pressure. Puberty increases levels and receptors of oxytocin (a bonding influence) and dopamine (incentive-processing); these are evidenced in teenage social behaviour.

Puberty itself is worth considering, as the surge in hormones that trigger it also create a need for social interaction, 'a neurobehavioural nudge' (Peper and Dahl 2013) towards negotiating and involving oneself in huge social changes. This is primarily down to the production of sex hormones. Positive or negative experiences here will influence an individual's entire life, in terms of social and affective learning and bias a person accordingly. Traditionally, testosterone has been blamed for high risk and aggressive behaviour in teenagers, but the real story appears to be more complex. According to Josephs, Metha and Carre among others (2011, in Peper and Dahl 2013) the social environment plays an important role not only in mediating the effects of testosterone on behaviour, but also influencing levels of the hormone. Testosterone in combination with other hormones may produce other effects—recent research has shown high testosterone levels combined with low levels of cortisol in boys exhibiting delinquent behaviour (Popma et al. 2007 in Peper and Dahl 2013). In girls, high levels of estradiol seem to be related to higher risk taking (Quinlan et al. 2010 in Peper and Dahl 2013). Goddings et al.'s research [Goddings et al., 2012 in Peper and Dahl 2013)] has placed DHEA, testosterone and estradiol within the anterior temporal lobe during 'socio-emotional processing'—this may be important as Zahn seems to have shown that guilt and shame (social emotions) are processed in the same place (Zahn et al. 2007 in Peper and Dahl 2013). Interactions between socio-affective processing and cognitive control 'heat up' considerably during puberty in the ventral striatum. If these interactions are positive, good things will ensure; if they are negative, and even if they begin in a small way, they may result in depression and substance abuse. These effects will not be 'one offs' they will 'cascade' across adolescence (Crone and Dahl 2012, in Peper and Dahl 2013). A lack of sleep too can produce tremendously negative socio-affective effects at this age (Carskadon 2011 in Peper and Dahl 2013). What is not known is the effect of space on human development in general and here hormonal development in particular—how might a lack of gravity affect the teenager? How might the regulation of light and dark affect sleeping patterns and circadian rhythms? How might practical problems such as these contribute to the social and emotional benefit of the adolescent and by extension the group? These issues will have to be planned for. Endocrinology is precariously balanced in the developing human to positive or negative effect—it may be necessary to control hormone development in space to ensure optimal human function. If the transhumanists have their way, this will not constitute an issue of liberty. If, however, by controlling hormonal production, I can alter a developing human's personality to benefit the group, how am I infringing that person's right (and freedom) to develop in the way their genes and environment might dictate? This is a more complex problem than just to control externally adolescent risk.

A possibly less complex solution, then, to games of 'chicken' in an already precarious off world context, might be to keep the teenagers busy in mixed age groups, not allow them to congregate, or to monitor congregation. At what stage of control might these steps constitute an infringement of their liberty? Should this possibility be negotiated or simply avoided? There will be a lot for adults in the group to decide, but there are plenty of sci-fi stories where the troublesome teenager is either eliminated or raised to the rank of revolutionary leader. Parents of course will be on hand—and will hopefully have enough time to devote to their offspring to help negotiate these difficulties.

Mothers in particular are the subject of some interesting current research in teen behaviour. In recent research by Gerardy et al. (2015), higher levels of 'consulting' between mother and teen seem to be indicators of pro-social behaviour. Where previous research focussed on indirect parental influences, it now appears that direct influence is beneficial—it results in better social adjustment and the development of further social skills (Gerardy et al. 2015). Some of this is dependent on the idea that the adolescent spends time away from the parental home in the company of peers, which might lead to the risky behaviour explored earlier and has been shown to result in negative socialising. It may not be possible in our context for teenagers to 'escape' their families; in fact in might be a very good idea to set up systems where they don't… Where and how might parents create, directly or indirectly, a supportive environment optimising teenage development?

Fathers in the literature have traditionally decreased the time spent with the children as the children get older. This may be because of a perception of threat, or because they are preoccupied with their own concerns. Fathers are more likely to try to exercise control over an adolescent, whereas a mother will be more likely to relinquish it. In father son relationships, there will be a decrease in a son's assertive behaviour as the father increases his own assertiveness. Mothers will still spend more time with their adolescents than fathers in an arrangement that seems to be mutually acceptable (Phares et al. 2009). In our context, where there may be less 'leaving' the domicile, it may be possible to change this pattern to the benefit of parent and teenager. Research seems to show that, even in the 21st century, fathers are less active in the parenting of adolescents than mothers.

Direct parental influence is divided into three categories by Ladd and Pettit (Ladd and Pettit 2002 in Gerardy et al. 2015). These are: consulting, guiding (mediating) and designing. It seems though that high levels of guiding, where there are too many rules, may have a negative effect on teens—sensitivity is needed here as young people need to develop autonomy in decision making. The Goldilocks amount then will require parents in our context who are willing to work hard at developing their (parenting) skills, balancing this with the discipline and determination necessary to survive in a hostile environment. Designing, which takes into account allowing setting and access to peers, must be handled in a similarly sensitive way. There will be the danger of a rise in negative parenting in such a refined environment off-world. Unedifying stories of the pioneers of the Wild West and some of the cruelty meted out to children in days past when they failed to perform as adults must be avoided absolutely in a space settlement of any kind. Negative or aggressive parenting will result in aggressive and anti-social children (Kawabata et al. 2011 in Gerardy et al. 2015) with the consequent risk to the group. Gerardy goes on to state that higher levels of maternal positive parenting are related to happier and better socially adjusted young people. Higher positive levels of consulting seem to result in positive emotions felt towards friends, so children experiencing this are going to develop better relationships. In our context, social cohesion and the effective management of emotion will be extremely important to the health of the group—maintaining levels of happiness will be crucial.

How then to keep teenagers happy while avoiding accusations of manipulation? The obvious idea here would be computer games. Edutainment—learning on a holodeck or in a simulated environment of some kind is already being touted as 'the way forward' by some in education. Computer games on board ship then—but just how satisfying might these be? How could they compare to learning an instrument (with its attendant neurological and psycho-social benefits as evidenced by El Sistema (Majno 2012) 'the system' of introducing orchestral instruments and playing to a disadvantaged community, resulting in improved social cohesion, learning, reduced depression and substance abuse among other things). If you are happy, you are unlikely to be a danger to your community. Recent research indicates, however, that the propensity for life-satisfaction in adolescents may have more to do with personality traits than the environment, so might it be possible to plan genetically for positive personality traits? If teens are slaves to their hormones and have taken a negative trajectory as explored above, they may lose the advantage of positivity in this regard and this may affect the group adversely. It may also affect or negate attempts to keep the group happy.

Recent research on a group of adolescents in Taipei (Chen 2008) suggests that their personality traits may indicate whether or not they consider online gaming to provide them with life-satisfaction. Online gaming is of course gaining in popularity all over the world; communities of gamers are multi-national, and games are played mostly in English, as to date translation devices are not yet sophisticated enough to function as quickly as might be required in a multi-lingual environment (although that day is coming). Gaming in an off-world context would not have the advantages it has on earth—no communication with significantly distant humans could occur, and they would be relying on the on-board computer or each other. However, it is worth exploring the idea that the adolescent personality and positive traits are important things, and how these might be influenced to maintain an optimal level for the benefit of the community. Interestingly, Chen's research appears to show that being extrovert and conscientious cannot be linked to greater satisfaction, but that being open, neurotic and agreeable in small measure may indeed impact on it. She suggests that designers might create games specifically tailored to gamers' personality traits (Chen 2008)—a concept explored in Neal Stephenson's The Diamond Age (Stephenson 1996) where an interactive book is designed to adapt to a young girl's circumstances to lead her through the best possible education and the best possible outcome for her. In this sense, such an approach might be too individualistic, as the success (and survival) of the group will depend on everyone's capacity to work together and not just on the best possible outcome for one individual. Considering Larry Niven's Ringworld (Niven 1970) also as a manifestation of individualism, it is Teela Brown's luck that takes her and the group to the Ringworld, where she finds the best possible environment for her own personal development, and everyone else is left 'dancing to the strings' of her luck for the rest of their days. Working on developing positive personality traits may involve endocrinological manipulation as well as activities designed to promote the joys of working and achieving together. In this case, given the proven success of El Sistema, we might consider sending an orchestra into space and teaching young people music. Without banging a drum (!), the transformational change in communities evidenced by this system is monumental (Manjo 2012) and bears further consideration. It certainly seems to achieve more than online gaming in terms of healthy social interaction, learning, brain development and social cohesion—all of which are more than desirable in an off world context.

To return to further psychological consideration of the Sturm und Drang of adolescents, Stanley Hall was the first to suggest that we were all to be doomed by the experience. This is of course not the case—we, the survivors of adolescence constitute the majority of the human population of earth and of course can guarantee the survival of our species thus far. But the numbers on earth are what counts—in an off world context, we could not afford to: lose anyone, risk a negative incursion into group cohesion or lose control of risky behaviour, especially when it is happening at a subconscious level. Casey and Caudle (2013) suggest that adolescence as an experience falls between the Stanley-Hall (1904) suggestion and the cultural (rather than biological) account given by Mead in her Coming of Age in Samoa of 1928. Casey and Caudle (2013) found that teenagers have in some instances better impulse control than adults or children, but that when emotion is added into the mix, this fails quicker than in adults and children. Therefore, if we want to put adolescents into positions where they have to make decisions quickly, it would be better to remove any emotional content from the decision-making environment. This is an important finding in terms of teenage function on board a ship or in a new-world context. Allowing them autonomy to develop while taxing them just enough to enable growth without rigid control mechanisms, encouraging them to negotiate challenges and push boundaries will be essential in the creation of fully rounded optimally functioning human beings equipped to carry forward the survival of the species.

Lastly, it is worth considering the issue of conflict resolution. The relationship between parent and child is immensely complex, and by the time a child reaches adolescence their level of secure attachment will be enabling them to cope or not with crises and to ask for help if they need it. Non securely attached children will have learned that whatever they do, no help will be forthcoming, so they are likely to give up more easily and not seek resolution. Approaching adolescence with these learned behaviours will prove pretty much disastrous. Parents who do not attach or attune to the children are modelling future misery. Van de Kolk (2014) relates the sad story of the 14 year who deliberately becomes a house-breaker over Christmas in order to have the police recognise who he is—spending Christmas in jail is worth it for hearing them call his name (Van de Kolk 2014). We humans really are quite simple creatures—we need other humans, we exist in relationship, we need to be loved. Who is the better rounded person in the recent film, The Martian? The PR woman, who doesn't recognise the term 'The Council of Elrond' or the scientists—the geeks? Where do they find a source of comfort? Which is likely to make a better parent? (Answers here are unnecessary, the film is full of gentle pokes at 'geek-dom'. Interestingly the astronauts and decision makers are not portrayed in this way, but the people responsible for the tech are.) How does this pertain to liberty? Is there not a collective responsibility to guarantee ship-born or off-world born children a right to pursue the 'good life' and to be equipped in the best possible way so to do by those responsible for their welfare?

Parental attention is necessary probably until we are 26 if we really want to function in the best way possible as adults. The concern in an off world context is that there will be so much to do and so many potential crises to overcome that children will be neglected and that their well-being will be compromised. Even babies in utero will feel and suffer from their mothers' traumatic experience (Van der Kolk 2014).The negative effects of this will manifest as aggression in the adolescent. In a study undertaken by Lyons-Ruth, 20 year olds with histories of seriously disrupted emotional communication patterns exhibited 'an unstable sense of self, self-damaging impulsivity, inappropriate and intense anger and recurrent suicidal behaviour' (Van der Kolk 2014, p. 120). Having an impaired sense of reality is associated with not feeling safe; this can start at a very early age-babies are aware of their own vulnerability and look for attachment with the mother; if this is not forthcoming, their personal emotional landscape is compromised. In space we would need to do our best to avoid this type of scenario, as the community will pay the price. The teenager who comes up fighting may do so for a number of different reasons, some more potentially toxic than others.

Aggression or conflict, then, may manifest itself for different reasons in our settlement or space ship, hopefully not for reasons to do with negative development or traumatic upbringing. If she is in a competitive play-off with someone else, there may be age-appropriate displays of temper, rather than physical combat. Rough play may offset some of the need for trials of strength. Whatever the manifestation, the appropriate thing to do will be to listen to both parties. It will be very important not to tell young people how to think or feel—drawing alongside and providing comfort may be more helpful. Affairs of the heart present us with very tricky manoeuvres—how to negotiate desire among the ship's teens if love is unrequited? Nearly everyone can remember a similar experience—but do we remember how we coped? Adolescence is a time of such intensity—and this can be a wonderful thing if we can help our children to negotiate it appropriately. Mindfulness, attunement and other practices can help restore the balance between the rational and emotional brains. Terror comes when you feel out of control—terror in an off world context will be more than super real; everyone will be on a knife edge knowing that the tiniest unforeseen event could bring annihilation. The stress of living like this will have traumatic effects on everyone, comparable perhaps to living in Belfast during the troubles of the 70s and 80s.

One of the most effective ways to deal with emotion, building resilience and intelligence while heaing, is to indulge in heightened play. Cicely Berry, voice coach for the RSC (and many others) successfully uses Shakespeare for this. In her work in the favelas of Rio de Janeiro, it is remarkable to see how disadvantaged teenagers are able to process negative emotions and feelings through, for example, Portia's speeches in the Merchant of Venice. On her DVD (Where Words Prevail, Berry 2005), teenagers perform their lived experience of rejection in the safe space that this heightened language provides. The benefits are clear—so does this leave us considering Shakespeare as a guarantor of emotional stability in an off-world context?

It seems then that successful negotiation of the Sturm und Drang period of adolescence in space may depend on the way we deploy our culture. This has huge ramifications for choices in terms of what we take on the trip—would it be possible to supply everyone with an instrument? The benefit of this in terms of social cohesion is currently unparalleled on earth, although it is very much a model of Western civilisation. Providing Shakespeare, Goethe, Schiller, Miller, Pinter, Wilde and other great playwrights of the world would not take up much storage space. But might these be enough to ensure smooth progress into adulthood, with or without an accompanying ritual, or might we need to rely on more physiological means? Could and should hormonal development be controlled to maximise ease of transition, could we accept this even if it is an infringement of liberty? To extrapolate this to a ridiculous degree, might it be possible to genetically manipulate a baby to ensure her development into the calm, competent, non-dangerous, intelligent and respectful human being who will function at her best under pressure at the helm?

As is usual in this series, I have tried to raise just some of the issues pertaining to humans, particularly children, in space, whether in the confines of a ship or on a new planet. Although it has been impossible to more than touch on many of the issues surrounding education, I believe it will be fundamental to the continuation of the human race to put in place carefully considered strategies for the well-being and development of children. We will not only need to consider what they will need to know, but how this knowledge should be delivered. The burden then on the adults will be huge—to function in the way they need to maintain a safe environment, but also to know how to teach. This is a skill not suited to everyone, and something I have not had the opportunity to address. How much are we planning to take for granted? How much would be fatal to assume? Considering that knowledge is power, leaving it to the onboard computer might be a mistake. Leaving the teenagers alone to mooch about in groups might also be an error. Careful management of the lives of the children, balancing their needs as individuals with the cohesion of the group will be essential. Let me hope that his chapter in some small way begins to unpack the amount of thinking that will be needed to guarantee life and liberty for future generations among the stars.

References

Albert, D., Chein, J., & Steinberg, L. (2013). The teenage brain: Peer influences on adolescent decision making. *Current Directions in Psychological Science, 22*(2), 114–120.

American Academy of Pediatricians' Committee on Adolescence. (2000). Suicide and suicide attempts in adolescents. *Pediatrics, 105*(4), 871–874.

Berry, C. (2005). *Where words prevail* (DVD). New York: Sojourner Media; Boston: Distributed by WGBH Boston Video, ©2005.

Casey, B. J. (2013a). The teenage brain—An overview. *Current Directions in Psychological Science, 22*(2), 80–81.

Casey, B. J. C. (2013b). The teenage brain: Self control. *Current Directions in Psychological Science, 22*(2), 82–87.

Carskadon, M. A. (2011). Sleep in adolescents: The perfect storm. *Pediatric Clinics of North America, 58*, 637–647.

Chen, L. S. L. (2008). Subjective well-being: Evidence from the different personality traits of online game teenager players. *Cyberpsychology and Behaviour, 11*(5).

Crone, E. A., & Dahl, R. E. (2012). Understanding adolescence as a period of social-affective engagement and goal flexibility. *Nature Reviews Neuroscience, 13*, 636–650.

Gardner, M., & Steinberg, L. (2005). Peer influence on risk taking, risk preference, and risky decision making in adolescence and adulthood: An experimental study. *Developmental Psychology, 41*, 625–635.

Gerardy, H., Mounts, N. S., Luckner, A. E., & Valentiner, D. P. (2015). Mothers' management of adolescent peer relationships: Associations with aggressive. *Prosocial, and Playful Behavior, The Journal of Genetic Psychology, 176*(5), 299–314.

Goddings, A. L., Heyes, S. B., Bird, G., Viner, R. M., & Blakemore, S. J. (2012). The relationship between puberty and social emotion processing. *Developmental Science, 15*, 801–811.

Greenfield, S. (2001). *Altered states of consciousness Social Research, 68*(3), 609–626.

Majno, M. (2012). From the model of El Sistema in Venezuela to current applications: Learning and integration through collective music education. *Annals of the New York Academy of Sciences: The Neurosciences and Music IV Learning and Memory*, 1252

Mead, M. (1928). *Coming of age in Samoa: A psychological study of primitive youth for Western civilisation*. New York: William Morrow and Company

Niven, L. (1970). *Ringworld*. New York: Ballantine Books

Parman, S. (1990). *Scottish Crofters: A historical ethnography of a celtic village fort worth*. Rinehart and Winston: Holt.

Peper, J. S., & Dahl, R. E. (2013). The teenage brain: Surging hormones—brain-behavior interactions during puberty. *Current Directions in Psychological Science, 22*(2), 134–139.

Phares, V., Fields, S., & Kamboukos, D. (2009). Fathers' and mothers' involvement with their adolescents. *Journal of Child and Family Studies*, *18*(1), 1–9.
Popma, A., Vermeiren, R., Geluk, C. A., Rinne, T., van den Brink, W., Knol, D. L., & Doreleijers, T. A. (2007). Cortisol moderates the relationship between testosterone and aggression in delinquent male adolescents. *Biological Psychiatry*, *61*, 405–411.
Quinlan, M. G., Duncan, A., Loiselle, C., Graffe, N., & Brake, W. G. (2010). Latent inhibition is affected by phase of estrous cycle in female rats. *Brain and Cognition*, *74*, 244–248.
Stanley-Hall, G. (1904). *Adolescence, its psychology and its relations to physiology, anthropology, sociology, sex, crime, religion, and education*, Vol. 2. New York: Appleton
Stephenson, N. (1996). *The diamond age*. London: Penguin Books
Van der Kolk, B. (2014). *The body keeps the score: Mind, brain and body in the transformation of trauma*. London: Allen Lane (Penguin)
Zahn, R., Moll, J., Krueger, F., Huey, E. D., Garrido, G., & Grafman, J. (2007). Social concepts are represented in the superior anterior temporal cortex. Proceedings of the National Academy of Sciences, USA, 104, 6430–6435

Chapter 12
What Happens to Dissent When Flight Is a Present Option?

Paul Rosenberg

Abstract It's something of a truism in physics that closed systems tend to entropy. If we contain a process—if we close it in, in some way—it will degrade faster than it normally would. And this principle clearly applies beyond physics. John B. Calhoun, for example, famously documented the elevated entropy of closed systems with rat populations Calhoun (Proc. Roy. Soc. Med. 66: 80–88, 1973). He gave his animals everything they could possibly need, but enclosed them in a limited space. Inside of their closed system, some males became aggressive, others withdrew psychologically, females stopped caring for their young, and eventually their population plummeted, even though they had plenty of food. Humans are not rats, of course: we are self-referential, thinking beings. Subjected to fully closed systems, we break down mentally. And it is precisely the entropy of closed human systems that underlies this chapter. I will be examining dissent: Cases of dissent where flight was an option, and cases of dissent where flight was not an option. The great difference in space, of course—once we have sufficient technology—is that flight is nearly always an option; on Earth, unfortunately, it is often not a realistic option.

Keywords Dissent · Genocide · Space · Exit taxes · Entropy · Rulership · Holocausts

12.1 *Geographically Open and Closed Systems*: Introduction

Before we look into the fate of dissenters in open systems (where flight is an option) and closed systems (where it is not an option), I'd like to establish the general nature of open and closed systems a bit further.

P. Rosenberg (✉)
29 South Webster Street, Suite 350, Naperville 60540, IL, USA
e-mail: p.rosenberg@cryptogroup.net
URL: http://www.freemansperspective.com

C.S. Cockell (ed.), *Dissent, Revolution and Liberty Beyond Earth*,
Space and Society, DOI 10.1007/978-3-319-29349-3_12

The nature of open and closed physical systems is very well understood, but applying the concept in other ways requires supplemental evidence. Just because an idea seems nice to us, doesn't meant that it is true.

The case I mentioned in the abstract to this article—the famous rat experiments of John B. Calhoun—provide some basis for the increased entropy of closed biological systems, even including some psychological component. To extend this principle to large human societies is certainly sensible, since humans are biological beings, and especially since Calhoun's experiments showed a psychological element; still, evidence is necessary. And, as it happens, there is such evidence, resting in the historical record that most of us are familiar with: the comparison of ancient Egypt to ancient Greece.

Egypt, as you can see in the map below, was a strongly contained society. The green areas on the map are the fertile areas of Egypt. All the rest is nearly-uninhabitable desert. There was, very literally, no safe path of escape for the vast majority of Egyptians. As a result, Egypt was a very static society, in which old ways were embraced and new ways shunned. While glorified in modern imaginations, Egypt was no glorious society; it was a primitive tyranny.

Egyptians farmers, for example, were tied to the land, and residence was restricted to immediate family members.[1] Egyptian society was highly stratified, and social status was carefully displayed. Farmers were a large majority of the population, but agricultural produce was directly owned by the state or a state affiliate (a temple or a noble family) (see Footnote 1). The Egyptian economy, or at least nearly all of it, was centrally organized and strictly controlled (see Footnote 1). Likewise Egyptian religion was monolithic. There was one religion, and little, if any, religious freedom. The exception of Akhenaten and his quasi-monotheism seemed to last no linger than his fairly short rulership.

Egypt was, in a word, static. It produced monuments to rulers and it grew grain. Aside from that, it innovated little… and very little considering its duration: from perhaps 3150 B.C. to (more or less) the Christian era (Fig. 12.1).

In contrast to Egypt stands the Greeks. Following their dark age (roughly 1200 B.C.–800 B.C.), they formed a massively decentralized culture, and one that afforded flight to any motivated person with a boat. Greek city-states were all stand-alone units, each with their own, independent modes of life. And, there were some one thousand of them. The map below shows only the major cities and colonies of about 550 B.C. That is, before Athens became the dominating city of the Aegean (Fig. 12.2).

Under the Egyptians, knowledge was substantially monopolized by the states. The intellectuals of the era—the priests—were keepers of the most important technical information of the time, astronomical knowledge. In the Greek Aegean, however, with its wildly decentralizing geography, no institution was able to assert control. None of them were able to monopolize knowledge.

[1]Manuelian (1998).

Fig. 12.1 Ancient Egypt, fertile areas highlighted (*credit* Jeff Dahl, Creative Commons Attribution-Share Alike)

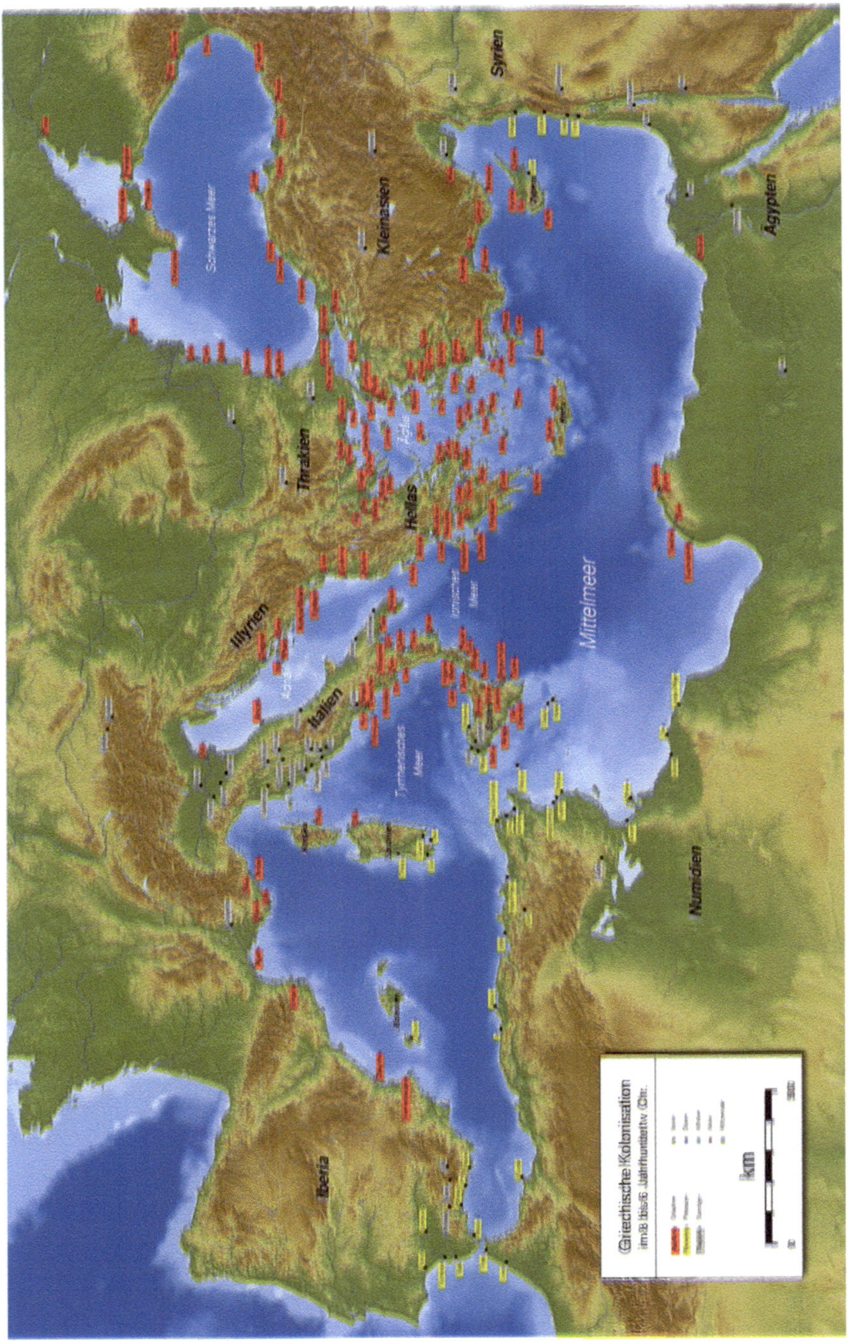

Fig. 12.2 Ancient Greece, showing the primary city-states (*credit* Maksim, Creative Commons Attribution-Share Alike)

In their condition as an open system, the men and women of ancient Greece gifted the world with literature, art, drama, geometry, astronomy, several kinds of philosophy, and much more. Had knowledge been contained inside of institutions, as in Egypt, had life been so very tightly controlled as under the Pharaohs, it is highly unlikely that the gifts we received from the Greeks would have formed. To borrow a phrase from Aristotle: they would not have enjoyed lives that afforded them the necessary scope.

There are other examples of this effect, all through human history. But I shall limit my examination to these two. Nonetheless, I believe that these examples, standing on top of the research of Calhoun and others, make it quite clear that closed human systems tend to entropy.

12.2 *When Flight Means Survival*: A Tale of Two Famines

To determine the value of flight (and I am here considering survival as a central value), it is necessary to compare one situation, where flight was an option, to another, where flight was not an option. For this comparison, I have chosen famines. And in particular, I have chosen the Irish Potato Famine and the Ukrainian Genocide Famine. The comparisons are not perfect, of course, but as we (fortunately) don't have a large number of well-documented famines to choose from, these seem to me to be the best comparators. So, as we proceed, I will point out the differences. One similarity between the two was that both had political causes, or at least major political contributions. Trying to quantify the importance of that similarity is beyond the scope of this article, but it seems worth noting.

Running for close to seven years, between 1845 and 1852, the Irish Potato Famine resulted in some one million deaths. Estimates vary from about 800,000–1.5 million, but one million seems a safe estimate.[2] I shall pass over the horrors of the famine to arrive at the significant factor in this analysis: It was possible for the Irish to flee this famine.

For a poor Irish peasant, flight was by no means easy, but it was, at least, possible. Immigration controls in those days were all but absent, and flight to North America was more or less unopposed. And the Irish did flee their famine. It seems rare that families fled as a unit, but very many families sent one or two at a time. And unlike most emigrations, women seem to have fled just as often as men, just as early, and in similar numbers.

It is further interesting to note that the Irish immigrants sent very significant amounts of money back to Ireland, most of which allowed other family members to follow them, and certainly increased the odds of survival for family members who remained. By 1851, these remittances reached £1,404,000.[3] This, then, is a second

[2]See Kennedy et al. (1999) and Gráda and Mokyr (1984).

[3]Foster (1988).

significant aspect to flight: It saves not just the person who flees, but helps others to survive as well.

In sharp contrast to the Irish famine stands the Ukranian famine. Running for less than two years, between 1932 and 1933, the Ukrainian famine was nearly impossible for a poor peasant to escape. The ideologues running the government of the USSR and their obedient enforcers saw to it.

The peasant farmers of the USSR did not have the right to leave the collective without permission.[4] Thus escape was all but impossible. Unsuccessful escapes resulted in immediate death or shipment to a gulag. The state further took control over the crops and paid the peasants whatever they liked. As a result, almost no one was able to flee the starvation. The result was massive death.

In a much shorter period than the Irish famine, some five million persons died. (Estimates run between 2.4[5] and 7.5 million.[6]) It can be argued that the Ukraine had a significantly larger population than did Ireland, but it is also true that large areas of the Ukraine (such as the more populous Western sectors) were spared, as shown in the map below (Fig. 12.3).

The chart below shows a simple comparison between these two genocides.

Famine	Duration	Deaths	Emigrants
Irish	6+ years	1 million	1.2 million
Ukrainian	Less than 2 years	5 million	Almost none

Lessons drawn from historical famines are—of a necessity—a long way from laboratory-controlled experiments, but neither are they to be passed-off. It seems highly likely that the death totals from the Irish famine would have been significantly higher—and perhaps double—if flight had not been an option. This provides us with at least a preliminary answer to our question of "what happens to dissent?" Clearly, when flight is an option, dissenters do much better. And when flight is not an option, dissent stands in grave jeopardy.

Our next section will make this point even more closely.

12.3 Further Genocides

The signature genocide of the twentieth century was the genocide of European Jews during the World War II era, and the Jews make a wonderful example of dissent. The Jews of Europe stood out, above all other reasons, because they dissented from the religion of the vast majority. But while Europe's Jews were very clearly a

[4]Fitzpatrick (1994).
[5]Snyder (2010).
[6]Marples (2007).

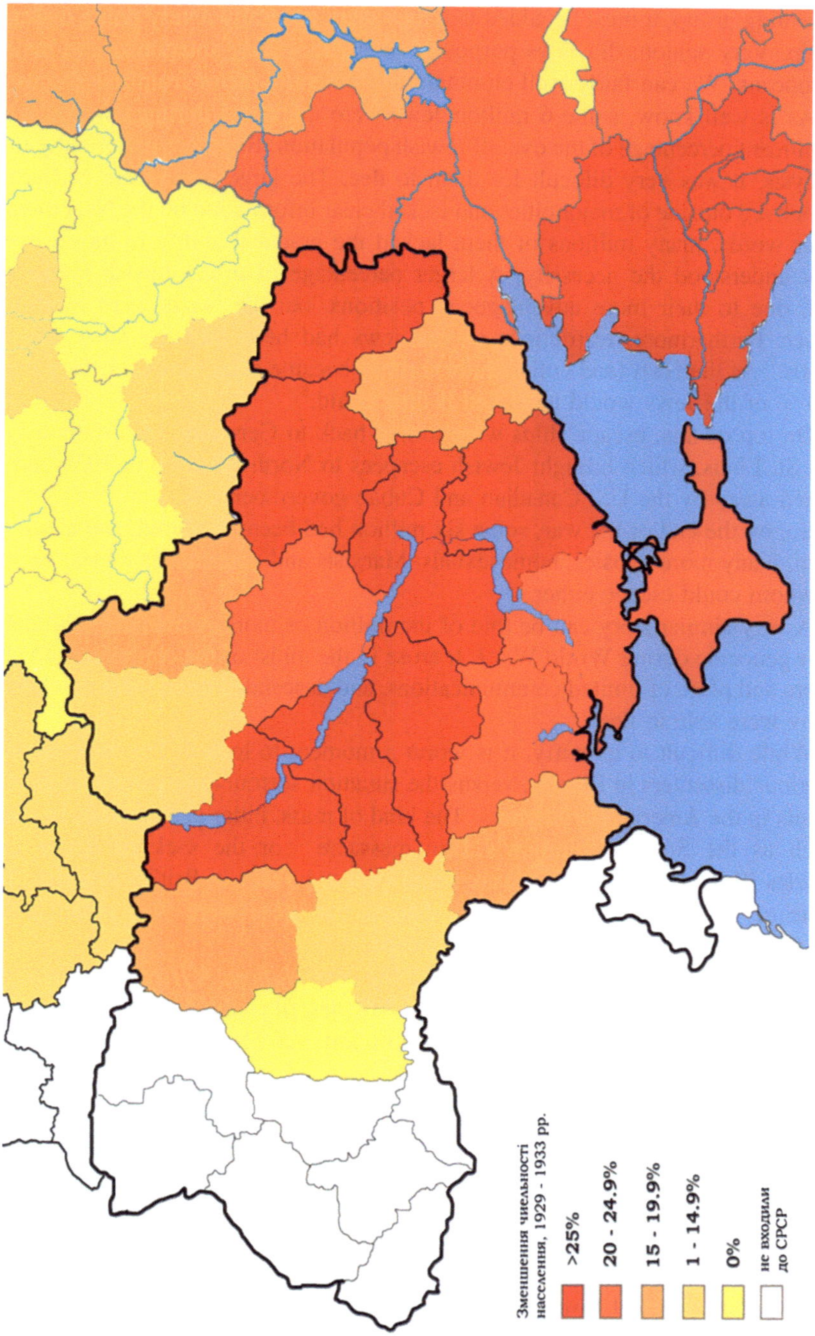

Fig. 12.3 Death rates in the Ukrainian Genocide famine (*credit* Sergento, Creative Commons Attribution-Share Alike)

dissenting group, it must be added that they were generally a very polite dissenting group. Very seldom did Jews purposely or flamboyantly antagonize the majority Europeans. We can fairly well eliminate that potential factor from this discussion.

As is well know, some 6 million Jews were murdered during this era; a very significant percentage of the overall Jewish population. But more importantly to this analysis, it was very difficult for them to flee. The large percentage of Europe's Jews lived outside of major cities and lacked clear information on the Nazi threat. In other words, many millions of them lacked the practical ability to move, even if they understood the necessity. A larger percentage of urban Jewish intellectuals fled, due to their more advantageous positions, but the Nazis were careful and clever. Furthermore, restrictions on the Jews had been building since the early 1930s.[7] As the Nazi (and collaborator) crimes became clear, however, very near to 100 % of the Jews would have fled if they could.[8]

On top of this, escape ships were turned back to Germany, most famously the MS St. Louis, which brought Jewish escapees to North America in 1939 and was turned away by the US, Canadian and Cuban governments.[9]

So, by the end of the war, some six million had been murdered, and along with them, many more gypsies, homosexuals, Marxists and other undesirables, very few of whom could escape either.

A very similar story can be told of the million or more Armenians who died in their genocide during World War I. It arose in the midst of the confusion of war, in an era and place of limited communications, and descended upon them quickly. Not many were able to flee.

While difficult to quantify, it is worth a moment to look at the persecutions of religious dissenters in Europe, versus the situation surrounding people of the same beliefs in the American wilderness. The kind of mass, bitter persecutions in Europe (such as the St. Bartholomew's Day massacre[10] or the sack of Rome under Charles III, Duke of Bourbon[11]) are simply not to be seen in British America. True, some of the puritans bloodied their hands, but these were very small actions, compared to those that transpired in Europe.

Historian Merrill Jensen, in his book, *The New Nation: A History of the United States During the Confederation, 1781–1789*, describes the conditions during this time, when not only was flight a very real option, but where people were dispersed in a wilderness:

[7] Friedländer (1997).

[8] Among other sources, I have verified these facts with several survivors of this holocaust in personal discussions.

[9] Morgan-Witts and Gordon (1994).

[10] Koenigsberger et al. (1999).

[11] Durant (1953).

> The back countrymen were of various religious sects, and while perhaps not tolerant by conviction, they were so on necessity on the frontier where so many religious groups had settled… All of these things [religious toleration and others] and more were expressed in petition after petition to colonial legislatures.

It is worth adding that many of the most horrific stories of history feature the sacking of cities. In most of those cases, the city was sacked after a siege… and a siege very clearly prevented flight.

Flight, as I think we have shown, is a crucial component to human survival during our all-too-common dark eras.

12.4 Is Flight More Readily Available in Space?

This, of course, brings us to a question: can we flee more easily in space?

In *The Meaning of Liberty Beyond Earth*[12] (Chap. 6), I argued that rulership is more or less doomed to irrelevance in space, at least over time. I explained that this is due to serval factors, but none more essential than simple physics: On earth, rulership must maintain a monopoly of force over a flat surface; in space, rulers must maintain the same monopoly on force in an ever-expanding three-dimensional area. The difference it that of a flat plane versus a sphere. The formula for the surface of a plane carries a square function; the formula for the area of a sphere carries a cube function. That is a very significant difference, especially given the difficulties governments have with two-dimensional space.

The Greeks, as mentioned above, were able to flee, so long as they had a sea-worthy vessel. Control of the Aegean, with its thousands of harbors and inlets, was simply impossible. Furthermore the 1000 greek city-states were fiercely independent—they did not enforce each other's laws and did not care about the other city's dissidents. The same and more will be true for space-faring humans, once space-worthy vessels are available. "Space-governments," however they may be imagined, will be hard-pressed to install and maintain control fully around every planet, moon and significant asteroid. Our galaxy is an extremely large place, with a diameter of some 900 Trillion miles.

As dissenters and humans in general are able to depart Earth, governments can hope only for control of the first few "hops." Beyond that point, the cube function required for three dimensional space dooms them… not to mention to immense distances involved. For anything resembling a foreseeable future, going a little farther will place anyone who wishes beyond the state's reach.

---

[12]Cockell et al. (2015).

12.5 Is Flight Permitted?

The problem with flight in our current circumstances is that flight is simply forbidden. This is true of more or less every government on the planet, and has been since passport treaties were created and enforced in the early part of the twentieth century.[13]

Flight, if we choose to admit it, is nearly always opposed by ruling regimes. None want their taxpayers (or in socialist countries, laborers) to leave. The USSR, for example, was notorious for allowing no one to leave. Emblematic in the modern world for a variation on the theme are the Exit Taxes that are imposed on people wishing to leave. The United States, Canada, Netherlands, Spain and South Africa impose such takings formally.[14]

In the face of such restrictions, entire industries of "offshore finance" and "legal expatriation" are thriving, even though the processes are difficult and often expensive. Dissenters desire the ability to flee, even when not being actively persecuted. Their opinions, it must be recognized, are legitimated by the work of Professor Rudolph Rummel, who calculated 262 million "deaths by government" occurring over the 20th century.[15]

The problem with flight, then, is that it is very often forbidden by the holders of power… also that these holders of power are also the abusers that one wishes to flee. As Dr. Martin Luther King explained in his *Letter from Birmingham Jail* (1963):

> We should never forget that everything Adolph Hitler did in Germany was "legal" and everything the Hungarian freedom fighters did in Hungary was "illegal."

When flight is a possibility, dissent survives.

12.6 The Virtue of Flight

The role of non-conformists in human history has been greatly under-appreciated. But it is appreciated by some, as Dr. Martin Luther King also wrote in *Strength to Love* (1963):

> Human salvation lies in the hands of the creatively maladjusted.

For reasons that are beyond the scope of this chapter, independently and uniquely arranging one's life—dissenting from the status quo—is regularly ridiculed and punished. Non-conformist failures are among the most favored stories for news broadcasts. People who see things differently are called "conspiracy theorists" and are thought to be somewhat deranged and dangerous. Conformity is thus

---

[13]Marrus (1985).

[14]See the Wikipedia page on "Expatriation tax".

[15]Rummel (1977).

enforced. A counter-idea, however—one that I call "the virtue of running away"—has a surprising level of historical support, which I will briefly explain:

Please consider this short list of people who "ran away" at one time or another in their lives, and often prior to their best work taking form:

- Abraham
- Moses
- Diogenes
- Pythagoras
- Sappho
- Buddha
- Jesus
- The Christian apostles
- Confucius
- Peter Abelard
- John of Salisbury
- Thomas Aquinas
- Stephan Langton
- Martin Luther
- Jan Hus
- Christopher Columbus
- John Locke
- Benjamin Franklin
- Thomas Paine
- Victor Hugo
- Daniel Defoe
- Frederic Chopin
- Leo Tolstoy
- John Dos Passos
- George Orwell
- Albert Einstein
- Nikola Tesla

Names could be added to this list and a few might be disputed, but I think the point is clear: It is very often the best of us who run away. Or, is it running away that makes some of us exceptional? Running away allows us to escape the web of expectations that almost invariably forms around us; it creates a separation that permits us to start afresh and re-create our lives.

Flight, however incompletely proven here, would seem to be an important factor in human development.

12.7 Conclusion: *Are Dissent and Flight the Path to Human Evolution?*

If, then, the ability to flee and actually running away are positive factors in human development, then dissent itself is a very important thing. However often dissenters choose their points of disagreement poorly (as has often happened), it would seem that the ability to choose, commit and separate from a status quo is, by itself, a significant factor in human development. Certainly it requires courage and certainly it involves an opening of mind. Both, it would seem, are important values.

I find a passage in Hannah Arendt's *The Origins of Totalitarianism* instructive in this area:

> The disturbing factor in the success of totalitarianism is the true selflessness of its adherents.

Regardless of the pop-psychology uses of "selfless" that surround us, selflessness is actually not a good thing. When we are truly selfless, as Arendt noted, we lose our capacity for individual judgment: we dare not judge for ourselves, we dare not stand on our own morals, we dare not differ from the crowd. We descend from being an independent producer of values to being merely an obedient cog. This form of existence makes us less conscious; less alive.

Furthermore, a life void of dissent—a life of mere compliance—weakens our civilizations. This damage is structural: The millions of units are prevented from direct production, and are forced to wait for instructions that move, slowly and unreliably, through a hierarchy. The result of this is denuded individuals and non-adaptive structures.

We are, at the final analysis, thinking creatures, and if we are prevented from thinking and acting according to our natures, we fail to thrive.

And so dissent, being action according to our individual natures, is crucial to human thriving, and should always be preserved by the ability to flee. Anything less is contrary to humanity's long-term survival.

References

Calhoun, J. B. (1973). Death squared: The explosive growth and demise of a mouse population. *Proceedings of Royal Society Medicine, 66*, 80–88.

Cockell, et al. (2015). *The meaning of liberty beyond earth*. Berlin: Springer.

Der Manuelian, P. (1998). *Egypt: The world of the pharaohs*. Bonner Straße, Cologne Germany: Könemann Verlagsgesellschaft mbH.

Durant, W. (1953). *The renaissance*. New York: Simon & Schuster.

Fitzpatrick, S. (1994). *Stalin's peasants: Resistance and survival in the Russian village after collectivization* (pp. 3–18). Oxford: Oxford University Press.

Foster, R. F. (1988). Modern Ireland 1600–1972. London: Penguin Group.

Friedländer, S. (1997). *The years of persecution: Nazi Germany and the Jews 1933–1939*. London: Weidenfeld & Nicolson.

Kennedy, L., Ell, P. S., Crawford, E. M., Clarkson, L. A. (1999). *Mapping the great Irish Famine*. Ireland: Four Courts Press.
Koenigsberger, H. G., Mosse, G. L., Bowler, G. Q. (1999). *Europe in the sixteenth century* (2nd Edn). Longman: Routledge.
Marples, D. R. (2007). *Heroes and villains: Creating national history in contemporary Ukraine* (p. 50). Hungary: Central European University Press.
Marrus, M. (1985). *The unwanted: European refugees in the twentieth century*. New York: Oxford University Press.
Morgan-Witts, M., Gordon, T. (1994). *Voyage of the damned* (2nd, revised (first in 1974) ed.). Stillwater, Minnesota: Motorbooks International.
Gráda, C. Ó., Mokyr, J., (1984) New developments in Irish population history 1700–1850. *Economic History Review*, *xxxvii*(4), 473–488.
Rummel, R. J. (1977). *Power kills: Democracy as a method of nonviolence*. New Brunswick: Transaction Books.
Snyder, T. (2010). *Bloodlands: Europe between Hitler and Stalin*. London: The Bodley Head.

Chapter 13
Cultural Evolution and the Colonial Imperative

Kelly C. Smith

Abstract Evolving populations can be described as influenced by two opposing forces: a conservative force preserving adaptations which suit the population to present circumstances and a radical one creating diversity in order to prepare for novel challenges. An excellent technique evolution uses to handle this tension is the creation of "stealth diversity," which does not affect the population until and unless it is adaptively useful. Colonies are a uniquely powerful way to do this, since they simultaneously eliminate the conservative influence of gene flow while posing new challenges that redirect selection. Turner's classic analysis of the interplay between the East coast and frontier cultures of the American west echoes these themes within the context of cultural evolution. As any colonial culture evolves, the mother culture attempts to exert a conservative influence, creating tension and reinforcing divergence. Nevertheless, if the colony survives, it can generate adaptive diversity that is as valuable to the mother culture as it was inaccessible to her on her own. Thus, American style democracy blossomed on the frontier, but eventually cross-pollinated with the mother culture and beyond, greatly enriching world culture. Since the conditions on an extraterrestrial colony will be fundamentally different from those shaping terrestrial ones, we can't predict their evolutionary trajectory in detail. However, the development of extraterrestrial colonies will certainly create additional adaptive diversity that could be critical to our long-term success as a species. Colonies should therefore be pursued, not as a means to particular cultural ends, but as experiments in cultural adaptation. Moreover, we must be open-minded about the fruits of these experiments, since our pretensions to universality are often little more than an instinctive conservatism that does not reflect the realities of colonial circumstances. Indeed, most arguments against the establishment of extraterrestrial colonies partake of either extremely short-term thinking or an unfounded degree of confidence in the capacity of terrestrial culture to adapt unaided. Therefore, any truly rational, long-term, strategic plan for humanity must include an aggressive program of extraterrestrial colonization.

K.C. Smith (✉)
Departments of Philosophy & Religion and Biological Sciences, Clemson University, Hardin Hall, Clemson, SC 29634, USA
e-mail: kcs@clemson.edu

C.S. Cockell (ed.), *Dissent, Revolution and Liberty Beyond Earth*,
Space and Society, DOI 10.1007/978-3-319-29349-3_13

Keywords Cultural evolution · Extraterrestrial colony · Universal cultural principles · Conservatism · Radicalism · Frederick Jackson Turner · Democracy · Cultural diversity · Adaptive landscape

13.1 Introductory Comments

Comparatively little serious academic work has been done on whether and how humanity should pursue colonies beyond Earth. Part of the reason for this is a simple lack of imagination—people tend not to worry about what may come until it actually arrives and certainly extraterrestrial colonies strike most academics as the stuff of science fiction. Yet I believe that future generations will likely look back on the next 100 years as the dawn of, in Stapledon's (1948) famous phrase, *interplanetary man*. NASA predicts that we will have evidence of life on other planets within the next 20 years, which will fire the popular imagination and fundamentally alter how we see our place in the universe. At the same time, our technology is rapidly approaching the point where long-term extraterrestrial colonies, at least on a small scale, will be feasible. It will thus not be long before we begin to think of ourselves as citizens of the universe.

We don't yet have the data to project in any detail how human colonization of space will proceed, much less what specific enrichments it will bring to our culture. However, it is possible to apply very general principles of evolutionary theory to elucidate some basic truths about the relationship between colonies and the long-term health of populations. We can thereby make a compelling case that extraterrestrial colonies are critical to the ultimate success of humanity. Of course this is a very complex issue and thus my goal is to create an evolutionary gestalt in favor of colonization, posing some very general questions rather than defending specific approaches or answers. The idea is not to examine individual trees or even the outlines of a forest, but to attempt to describe the conditions needed for forests in general.

All populations and cultures must generate and maintain adaptive diversity lest they fail to meet the challenges changing circumstance will inevitably bring to their door. Yet such diversity always threatens the status quo and therefore any evolving population can be described as being driven by a dynamic balance between conservative and radical forces. The former seeks to preserve the adaptations that proved useful in the past, while the latter seeks novel adaptations that may hold the keys to future challenges. Colonies are a powerful tool in the service of novelty, since they allow an established population to explore adaptive possibilities without upsetting the status quo. As Turner (2013) described in the case of the American frontier, there is a predictable dynamic between the mother culture and the colony which can produce counterproductive tensions. However, the changes colonies bring can also revitalize a culture, and thus we should adopt colonization of other planets as a long term strategic goal. But we must also learn to overcome our instinctive conservatism to minimize friction and maximize cultural benefit.

13.2 The Evolution of Diversity

13.2.1 Conservatives Versus Radicals

Evolution is just a special kind of sorting process that will operate on any suitably situated population. Any trait within a population will tend to evolve by natural selection if:

1. It varies within the population: not everyone has the trait.
2. It is heritable: children resemble their parents w.r.t. the trait.
3. It is adaptive: the trait impacts the ability of its possessors to survive and reproduce.

The classic examples of evolution used by Darwin are all biological, but evolution by natural selection will operate in any population if these three conditions are met, whether they are populations of integrated circuits or ideas. Thus it's possible to model the evolution of culture using these three elements, though some of the specific dynamics will differ between biological and cultural situations since, for example, the way in which ideas are transmitted is not quite the same as with genes (Cavalli-Sforza and Feldman 1981; Mesoudi and Whiten 2008).

Evolution is a process fueled by diversity. If a population does not exhibit variance in a trait, it can't evolve. It's even been suggested that this is one of the reasons biology has a fundamentally different character than the natural sciences (and, in particular, why it lacks universal laws of nature) (Boyd 1991; Rosenberg 2007; Smith 2015c). So on the one hand, a population needs new diversity in order to effectively adapt to changing conditions. On the other hand, such diversity tends to compete with and dilute the effectiveness of existing adaptations that suit the population to its current conditions. Diversity is thus a double-edged evolutionary sword, as it is both the key to the future and a threat to the present.

Given this dual nature, it is not surprising that evolving populations experience competing "forces" that push in opposite directions, producing a dynamic resultant. If we indulge in a bit of anthropomorphism and give voice to the conservative tendency, it might say, "There are good reasons for things being just as they are. So even when novelty seems tempting, you are well advised to stay the course." But populations also harbor a radical force that would respond with, "The world is changing and we must change with it. The old ways don't seem able to handle these new situations and much hinges on our ability to embrace change."

The conservative force will tend to be stronger whenever environmental change is minimal. In such circumstances, it's a good bet that traits put in place by past natural selection will continue to deal successfully with the familiar challenges imposed by the environment. Thus in biology, morphological stasis over long periods of time, as with horseshoe crabs, is taken as evidence of unusually stable environmental conditions (Sekiguchi and Shuster 2009). Highly traditional cultures are often found in such circumstances as well, as happened with China after it created an unusually stable cultural environment by adopting extreme isolationist policies in the 15th century (Landes 1998).

Beyond environmental stasis, the primary force in favor of conservative forces is "gene flow." Any new variant (gene or idea), must struggle to gain traction within the population at large. Since it begins its journey with very low incidence in the population, this is a difficult task and the outcome is far from certain, even if the newcomer confers a clear adaptive advantage. The larger the population and the more genes or ideas mix freely, the more competition a novel variant will have to overcome. Therefore, as a general rule, conservatism is stronger in large populations with lots of gene flow.

Contrariwise, a changing environment and a small population favor the radical force. The faster and more dramatically the environmental challenges to a population change in character, the less likely the adaptations favored by past natural selection will be effective in dealing with them. Indeed, change tends to be self-reinforcing, since as a population responds to novel challenges by evolving, this alters the competitive matrix within the population to create further challenges. This frenetic state of affairs is captured by van Valen's (1973) famous "red queen hypothesis", which proposes that organisms must constantly adapt just to *maintain* their relative position within a population of other adapting organisms. As the red queen herself puts it:

> Now, here, you see, it takes all the running you can do, to keep in the same place. If you want to get somewhere else, you must run at least twice as fast as that! (Carroll 2014, p. 128).

Small populations, with their smaller pools of genes or ideas, also favor change. A single copy of a novel variant in a small population will begin its trek at a higher percentage of the total diversity and thus have less distance to travel to become established. Small populations are also more likely to contain a biased sample of the original population's adaptive variance, and thus may be missing some of the greatest challengers the newcomer would otherwise have to outcompete. As a general rule, therefore, small populations and those with restricted gene flow tend to evolve much more rapidly than large ones.

Evolution must strike a dynamic balance between these two forces. It can not allow too much novelty in the population lest this disrupt the hard won gains of past evolution. Yet no matter how well things are going at present, change will eventually come and thus it must also find a way to foster and maintain adaptive diversity. What would be ideal is if a population could maintain "stealth diversity"—that is, diversity that does not make its presence felt until and unless it is needed.

13.2.2 Managing Diversity

There are a number of evolutionary mechanisms that produce something like stealth diversity. In genetics, it is difficult to alter a gene that codes for a truly essential trait, since the most likely result of a mutation would be loss of function, followed by death. But errors in gene replication sometimes result in two copies of the essential gene alongside one another. Since one of the genes if now functionally redundant, it is free to evolve in new directions without cost. Duplicate genes can serve as a pool of stealth diversity with little effect on the organism until and unless they come under the influence of selection. As a result, we often see families of related genes with similar but distinct functions, as in the case of human hemoglobin (Efstratiadis et al. 1980).

Another very common technique for managing diversity is through gametes or juveniles. In many organisms, juveniles are relatively cheap to produce and thus the loss of any one is not a major problem. As a consequence, juveniles are much freer to explore potentially adaptive diversity without major risk to the parent population. For example, the physical dispersal of young is one way to explore the adaptive potential of surrounding areas without exposing the parent organism to risk. A tree that invests in widespread seed dispersal is remotely probing the local environment for more fertile conditions. Most of the seeds die, of course, but occasionally a seed will happen upon an advantageous setting and take root. Humans and other primate cultures often use juveniles, particularly young males, to explore new situations and ideas (Wilson and Daley 1985). To facilitate this, humans are typically much more receptive to new ideas and ways of doing things when they are young, becoming more conservative with age (Henningham 1996; Maltby 1997; Grant et al. 2001). Often, these youthful forays come to naught, but sometimes they discover something of great value. It's thus been argued that it's no coincidence scientists at the forefront of revolutionary development are often relatively young, as this makes them are more willing to explore ideas beyond the dominant paradigm (Kuhn 2012; Dyson 2014).[1]

One extremely effective way of creating and maintaining stealth diversity, whether cultural or biological, is through colonies. In its simplest form, a colony is a splinter of the original population that is: (1) separate from its parent in terms of gene flow, and (2) exposed to a new environment that poses novel adaptive challenges. The seperation between parent and colony allows the colony to evolve without disrupting the adapative status quo, much as with a duplicated gene. And even the complete loss of the colony will typically impose only modest resources costs on the parent population. The new adaptive challenges, coupled with weakened gene flow between the colony and the parent population, will cause the colony to evolve very rapidly. Consider, for example, the Galapagos finches Darwin made famous. Lying about 1000 km off the coast of Ecuador, the Galapagos Islands are very difficult to reach from the mainland.

[1]There are also cultures that have institutionalized this function and employ a small percentage of adults for exploration—for example, the ancient Polynesians bestowed high status on the adventurous souls who helped them spread across the Pacific islands (Irwin 1989).

But, very occasionally, circumstances will conspire to found a colony there, as when a pregnant bird is blown off course by a storm. The resulting population is cut off from the mother population and subject to very different selective conditions, so it evolves rapidly. Darwin initially took the diverse types of birds he found there to be separate species and was amazed to learn that they were in fact just highly distinct varieties of finch. The rapid evolution of the Galapagos colonies allowed Darwin to see the process of evolution in mid stride, as it were.

13.3 Dynamics of Human Colonies

13.3.1 The Adaptive Landscape

Now I'd like to introduce a visualization tool commonly employed in evolutionary biology: the "adaptive landscape." The adaptive landscape is an excellent way to intuitively grasp the tradeoffs between radical and conservative forces in an evolving population and so is well suited for our purposes. Populations are represented by balls rolling, under the influence of natural selection, on a landscape of potential adaptations. The topology of the landscape is determined by a complex array of environmental factors, much as massive objects are sometimes portrayed as deforming the fabric of space to create gravity wells.[2]

If we think of physical balls on a plane surface, we understand how gravity will act on each ball, propelling it across the landscape. Gravity will "seek" the lowest energy state possible, though the contours of the landscape and the initial starting position affect what is possible. So while a ball will roll downhill, in a complex landscape it will probably enter a basin of attraction that, while lower in energy than its local surroundings, is nevertheless not the lowest energy in an overall or global sense. At that point, it will likely get stuck, since leaving this basin requires that it go against the force of gravity and it has no motive force of its own. On an adaptive landscape, each population will roll downwards, "seeking" the highest fitness accessible to them. Selection will push populations across the landscape of adaptive challenges and opportunities towards basins of attraction that represent local fitness optima. Populations that attain a local optimum will likely get stuck, since they would have to go against the power of natural selection to continue their journey.[3] Thus, while we can confidently predict that a given population will increase in

[2]Here again, I must admit to major simplifications for the purpose of illustration. The adaptive landscape is already an idealized model, and my version is further simplified so as to be almost a caricature. In particular, it: (1) is two dimensional, representing natural selection as acting to optimize only a single trait, (2) assumes the landscape is static over a long period of time, and (3) treats "democracy" as a simple trait with, for example, a simple and unidirectional impact on the fitness of populations expressing it.

[3]There are other forces besides natural selection that act on populations, so they have some ability to move on their own. However, such forces are typically weak in comparison to selection and thus will often simply bounce the population back and forth at the bottom of a local optimum.

fitness over time, most will have to settle for less than perfect optimality. In terms of our earlier discussion, the radical forces hold sway as long as the population is rolling across the landscape, but once in a local optima, conservative forces exert themselves to squelch further adaptive exploration.

13.3.2 *Turner's Frontier Thesis*

Let's examine the founding of America by Britain to see this evolutionary dynamic in a cultural context. The right side of Fig. 13.1 shows two populations representing Britain and France in the age of colonization, each occupying a local optimum. If we assume that democratic ideas correlate well with cultural fitness and that Britain was at the time more democratic in an important sense, then the British population occupied a local optimum that was more fit in an overall or global sense than its neighbor. However, each culture is well adapted to its local conditions: in other words, there are probably good reasons for many of the features of both French and British culture, including their distinctive approaches to democracy.

Evolution is not blind, but it is quite shortsighted, since it can't see what opportunities may exist beyond the barriers defining a local optimum. From the limited perspective of a population at the bottom of an optimum, the immediate effect of change will be to reduce rather than enhance fitness. Therefore, in both populations, conservative forces will work to oppose exploration of the surrounding

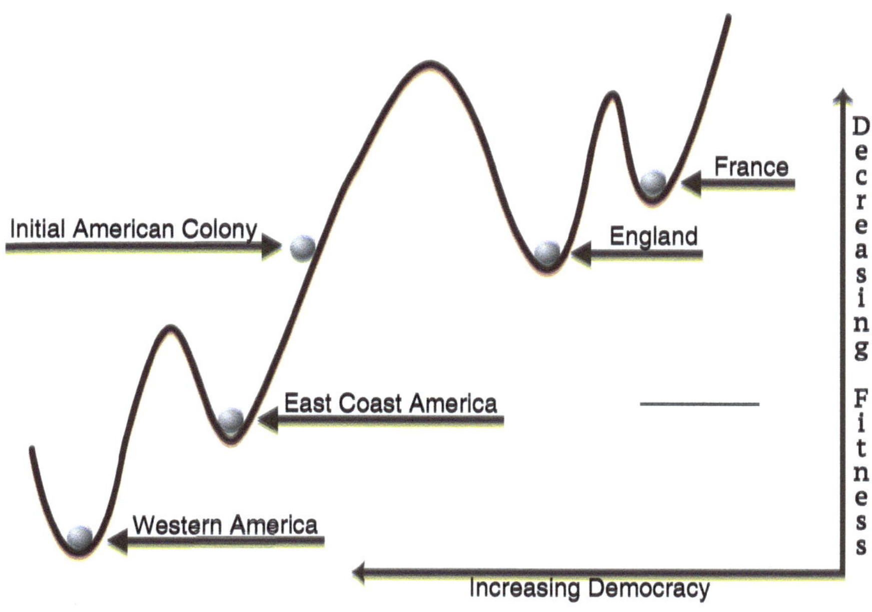

Fig. 13.1 Founding of America by Britain—evolutionary dynamics

landscape—a situation visually represented by the steep fitness peaks hemming in the two cultures. The shape of these peaks is the result of a number of environmental variables, including cultural practices. For example, current French law requires the use of the French language in all commercial and workplace interactions—a policy purposefully designed as a barrier to the diffusion of Anglo-American culture (Määttä 2005). In our simplified model, even if the French realized that they were globally less fit and wished to emulate British democracy, they would have an extremely difficult time overcoming the fitness barrier separating the two cultures. The French are thus trapped in their Frenchness.

As we track left on Fig. 13.1, we move ahead in time and cross the Atlantic ocean to the ball representing the early British colonies in North America. In terms of democratic traits, these begin at roughly the same point as mother Britain, but the selective environment of the new world, and thus the shape of the local adaptive landscape, is very different. In particular, the ability of mother Britain to exert its traditional conservative influence on cultural change is steeply curtailed by the intervening Atlantic ocean and other factors. There is thus no longer a steep fitness barrier to further democratization and so selection will begin to push the American population towards a new, more democratic, optimum.

Of course, the mother culture's actions are still driven by conservative forces, despite the difficulties of applying them to the colonies. These forces are not perfectly rational, or more accurately, they are rational responses to the adaptive landscape of the mother culture, but not necessarily to that of the colony, which is quite different. If societies were perfectly rational, they could calibrate their responses to different circumstances in a nimble fashion, but they are not. The British are every bit as stuck in their Britishness as the French in their Frenchness. From the point of view of individual members of British society, this conservatism plays out as a belief in the *objective* superiority of their own cultural practices, with predictable increases in the coefficient of lifted noses and use of derogatory terms like "uncultured," and "barbarian" to refer to the colonists. From their perspective, the friction between colony and mother culture is the fault of the colonists, who don't understand what's good for them (British culture). Of course, the colonists have a different perspective. They see the cultural changes they embody as perfectly appropriate responses to local conditions and identify the source of the friction as British ignorance combined with arrogance: even though the British have little experience in the new world, they are still supremely confident they know what's best for those who live there.

The result is a cultural feedback loop. As the colonies become more distinct, they encounter more conservative opposition from the mother culture. This in turn leads to friction and resentment on both sides, which accelerates cultural differentiation and steepens the fitness barrier separating the two cultures, locking the differences in place. This dynamic is captured in the sort of scene one often sees in fictional portrayals of the era, where colonials meet in a tavern to express their grievances:

> Those damn English snobs! They've never skinned a beaver or fought an Indian, yet they see fit to tell us how to run our affairs. They treat us like ignorant children and refuse to let us vote on the laws that govern us. This is our land!

When this dynamic proceeds far enough and other conditions permit, the colony breaks away to become a new society, as happened with the American revolution.

But for America, the story does not end there. As the historian Frederick Jackson Turner (2013) famously argued, the continent of North America represented an opportunity perhaps unprecedented in recorded history. The vast frontier between American's coasts made possible the continuation of the colonial dynamic for an unusually long period. The "East coast" population in Fig. 13.1 has already adapted to local conditions, finding a more egalitarian and democratic optima than mother Britain and insuring these are well protected from further British influence. But before it stretches a slope representing 100 years of continual frontier exploration and the adaptive potential this promises. In evolutionary terms, the American frontier is just a series of colonies moving further and further Westward, away from the east coast establishment. As these successive waves of colonists move Westward, they both encounter novel adaptive challenges and insulate themselves from the conservative influence of the East Coast. As with any colony, this results in rapid cultural evolution, in this case eventually producing a uniquely egalitarian brand of American democracy.

The fact that the East coast establishment was recently a colony chaffing under British rule did little to prevent it acting paternalistically towards its own colonies, as any mother culture tends to do. Just as were the British before them, those on the East coast were both ignorant of frontier conditions and confident that their own practices were objectively best. Imagine a group of cowboys gathered around a campfire out on the range, giving vent to their frustrations:

> Those damn city slickers back East! They've never ridden the range or branded a cow, yet they see fit to tell us how to run our business. They treat us like ignorant children and refuse to grant us statehood. This is our land!

Of course, America managed to reintegrate the two cultures relatively peacefully, in large part because new transportation and communication technologies reestablished continent-wide gene flow relatively quickly.[4]

Turner and others have argued that the ideas born on the American frontier migrated back to enrich not just American society, but world culture in general and indeed was an important element in the founding of our current democratic age (Rosenberg 2011; Turner 2013). Britain's investment in the American colony produced cultural diversity that created initial friction with their own established values, but ultimately proved an investment that paid rich dividends.

I certainly don't want to imply that change is not possible without colonies. Mother cultures are not entirely monolithic or unchanging, though such terms are

---

[4]However, it could be argued that one of the many causes of the American civil war was lingering resentment between the urbanized East and the more rural South, which echoed this earlier dynamic.

relative. The question is not *whether* a mother culture can change in the face of conservative forces—it certainly can. Rather, it's how *much* it can change and how *rapidly*. It will usually take much longer for an established culture to make changes than a colony. In biology, for example, while the creation of a new species is possible in the face of gene flow (sympatric speciation) it is much more difficult (Via 2001). In general, therefore, colonies evolve faster though this doesn't always produce good results since it's also easier for a colony to evolve in *maladaptive* ways, which is why mature cultures have conservative forces in the first place.

My account is certainly a major simplification of the historical and cultural situation. For example, Turner's ideas have been critiqued on a number of grounds: that frontiers are ever present, that factors other than democracy were important, that cities were more central to the story than Turner allows, and that he underestimates the contributions of women (Pierson 1942; Elkins and McKitrick 1954; Limerick 1987; Riley 1993; Wade 1996). In terms of my argument, however, these amount to debates about the precise forces that shape the landscape and the exact topology their influence produces, and do not affect the big picture, whose general conclusions remain secure, namely:

1. Cultures tend to become stuck in local optima due to conservative forces which are usually only locally optimal.
2. This hobbles the creation of the diversity that can be critical to the long-term ability of a culture to meet new challenges.
3. Colonies help overcome this dilemma by evolving rapidly and independently.
4. Tensions between the mother culture and colony arise and self-reinforce, fostering further differentiation.
5. The resulting colonial adaptations can then influence the mother culture in positive ways.

Of course, the details of this process can and no doubt will be debated endlessly, but for all its imperfections, Turner's account serves as a useful illustration of some aspects of colony-parent interactions that are likely universal.

13.3.3 Extraterrestrial Colonies

There's every reason to believe that extraterrestrial colonies would follow much the same general pattern we have seen with colonies on Earth. Of course, at present, the barriers separating terrestrial culture from extraterrestrial colonies are prohibitively steep because we lack the technology to make colonies practical. I will leave it to others far more qualified than I to discuss when and how we might be able to overcome these hurdles, and restrict myself to a very general discussion of the advantages that would accrue to humanity once we do.

An extraterrestrial colony would certainly be quite isolated from the influence of the mother culture on Earth. Even colonies within the solar system would be roughly as difficult to reach and administer as the far-flung colonies established by the Europeans in the age of sail. Any colonies established in other systems would be essentially impossible to administer remotely and thus would have to be self sufficient from the very beginning. Evolutionary theory thus predicts that extraterrestrial colonies will undergo divergent cultural selection and rapidly distance themselves from the mother culture, as has happened many times before.

Extraterrestrial colonies would also evolve in a very different adaptive landscape than terrestrial cultures we are familiar with. Cockell (2002) has even argued that the conditions would be so different that use of Turner's frontier thesis is problematic in part because extraterrestrial colonies might veer away from democratic government. He points out that any extraterrestrial colony within our own solar system would have to manufacture its own oxygen and water and this infrastructure would have to be constructed and controlled centrally, in which case the danger of totalitarian rule looms large—he who controls the oxygen, controls the population. This is an excellent example of how the *details* of extraterrestrial colonies, and thus the adaptive landscape, would be different from previous colonies on Earth. But we have to be careful what lesson to draw from this difference. Such examples should be sobering to anyone who takes from Turner the lesson that colonization always favors democracy. But our predictions need not be so specific. If all we claim is that colonies will tend to rapidly evolve to face new challenges, then the precise cultural trajectory—whether towards democracy or totalitarianism—is not critical. While the American frontier strengthened democratic values, other colonies may evolve in other directions. Indeed, the observation that space colonies may evolve in new ways that representatives of terrestrial culture don't like just makes the larger point: colonies will explore new cultural territory and, in so doing, adapt in ways that generate a conservative backlash from the establishment.

13.4 Universal Cultural Principles

13.4.1 Extraterrestrial Culture

Specific predictions about evolutionary trajectories are always hard to justify, but this is especially true in the case of extraterrestrial colonies since this is virgin territory. However, let's assume for the purposes of argument that totalitarianism is a basin of attraction towards which space colonies will tend to evolve. The question then becomes, "To what extent are we justified in resisting this change?" Put another way; to what extent are we confident that our distaste for totalitarian governments represents a universal as opposed to merely a local optimum? Of course, as representatives of a mature democratic culture, we all experience a visceral reaction to any suggestion of totalitarian rule. But a large part of the reason for this is simply conservative instinct, of the same sort that produced a visceral reaction in the British towards their American cousins' experiment with "mob rule." Because the present conditions of our own culture make totalitarian systems highly problematic, we tend to think of totalitarianism as being imposed on unwilling subjects whose rights are trampled in the process. But it's not clear that a democratic government is always *possible*, much less desirable under the kinds of conditions extraterrestrial colonies may encounter.

I certainly don't want to defend totalitarianism in a universal sense—in fact, I would argue that liberal societies are probably better able in general to foster and evaluate cultural diversity than repressive ones. But it's worth asking the skeptical question suggested by the cultural dynamics we have been exploring. After all, totalitarian governments have been a common feature of terrestrial societies when conditions were very different than they are today. It is simply not obvious that democratic ideas *could* have taken root earlier than they did, much less that, if they had, we would all have been better off. Since we can't directly experience the conditions that lead to authoritarian rule in times past, we must rely on intuitions shaped by a different set of experiences. In other words, when we examine the distant past, we are in the same conservative position as a mother culture looking down its nose at its rambunctious progeny. If we are truly honest with ourselves, we have to grant that it's an open question whether more liberal forms of government are better in some universal sense.

In the case of an extraterrestrial colony, we can easily imagine circumstances in which a totalitarian government is not *imposed* on a colony but instead demanded by the colonists themselves. Imagine there were those who, for whatever reason, wished to see the colony fail—indeed, so desperately that they were willing to give their lives to the cause. A small colony would be highly vulnerable to attack: a simple metal rod launched from a ship in orbit could easily wipe it out entirely (Adams 2006) and a colony of any size that was dependent on the complex, centralized, and fragile infrastructure needed to produce water and oxygen would be highly vulnerable to sabotage. If you thought it a simple matter for someone to, without warning, destroy everyone and everything you cared about, what would you do? Recent history seems to indicate that people are in fact quite willing to give up hard won freedoms when they feel sufficiently threatened, even in a mature democratic culture. So it seems quite possible that colonists in such conditions would *demand* repressive measures such as enforcing a strict "no orbit" policy, militarizing elements of society, imposing harsh immigration controls, and pursuing invasive surveillance of the population. In other words, the very same aspects of a colony that make it worryingly easy to exert totalitarian control could also make such control *adaptive*.

Of course, there may be more democratic ways to handle such challenges and, if so, they are well worth exploring.[5] But then again, there may not be. And even if the cultural practices we hold dear can meet this particular challenge, it's a safe bet that there will be other challenges we can't see clearly from the bottom of our local optimum that will require other adaptations we won't like. After all, we have the attitudes we do in large measure because there are good local reasons to avoid such

[5]Cockell allows that some these sorts of measures might be needed, but hopes that they could be done in a relatively transparent and open fashion (Smith 2015b).

practices on present day Earth. But this is precisely why colonies are so important. A colony allows a mature conservative culture to explore new types of diversity beyond the fitness barriers of its own local optimum—diversity that may ultimately hold the key to its own long-term growth and survival.

13.4.2 Prescriptive Humility

It's easy to envision other potential assaults on extraterrestrial culture from terrestrial conservatives. Consider Stapledon's (1948) suggestion that humankind might wish to genetically engineering varieties of humans adapted to the harsh conditions of other worlds in order to facilitate colonization. Should others take up this call, terrestrial ethicists will no doubt wax eloquent in their moral opposition, as they have already done with much more limited applications of genetic engineering (Rollin 1995; Schurman and Munro 2006). Similarly, environmental ethicists will likely condemn entire categories of resource "exploitation" by way of analogy to terrestrial contexts.[6] Some of these reactions may ultimately be justified, but all too often they will be accepted simply because they appeal to our terrestrial intuitions. We thus need to be reflective about our conservative bias and do our best to evaluate extraterrestrial practices within extraterrestrial contexts.

For example, we are used to thinking of strip mining as a bad idea for any number of reasons that are compelling on Earth, such as pollution and destruction of habitats. But the situation would be quite different on the moon. For one thing, there is no biosphere to impact through our actions and thus no habitats to be destroyed. It's likely that the entire concept of "pollution" will need to be reconsidered: Is it pollution to move lunar regolith from one place to another in a way that has no direct negative impact on anyone? Of course, terrestrial philosophers can, and no doubt will, debate these issues endlessly. But whatever they conclude, it seems likely that colonists on other worlds won't be able to *afford* an environmental ethic suited to a rich and mature society on a living planet. Mere survival under conditions harsher than any human being on Earth has ever experienced will be enormously challenging. Colonists will thus be forced to use whatever resources they have to hand in whatever way they can to deal with the harsh realities they face. In that case, the question we should be asking is not whether it's moral to establish "environmentally unfriendly" colonies, but whether we should establish colonies at all, since "environmentally friendly extraterrestrial colony" may be an oxymoron. That is a legitimate question, to be sure—but a very different one than *how* best to set up a colony.

Of course, I don't mean to imply that *whatever* practice a colony adopts is morally defensible or that the fact the colonists *believe* their actions are justified means that

[6]At a recent astrobiology ethics conference, a debate erupted concerning the permissibility of strip-mining the dark side of the moon (Smith 2015a).

they *are*. Colonists could be mistaken about the fitness of these practices in either a local or a global sense. But this truth cuts both ways, since if we endorse this argument, we must acknowledge the implications for our own attitudes. As a mature culture, it's probably safe to assume that we have explored the contours of our *local* optimum well, so our confidence that these practices are well adapted to *local* conditions is probably justified. But this is a far cry from justifying these practices as universal in the sense that they should be adopted by *all* human societies, regardless of conditions. However diverse the conditions on Earth have been, they are not as diverse as those we will encounter when we leave Earth behind and, however brilliant the philosopher, she can not exceed the bounds of her imagination.[7]

So while we can have grounds to critique the practices of other societies, I think we must cultivate a strong sense of prescriptive humility. We would do well to keep in mind that much of terrestrial imperialism was justified by reference to ethical principles considered universal by the imperialists. Thus no less a moral luminary than John Stuart Mill argued that imperialist policies in places like India helped the indigenous people because Britain "*...of all in existence, best understands liberty.*" (Mill 1977, p. 565).

At the very least, our confidence in the correctness of our cultural prescriptions should decrease as the gap between the conditions of our own culture and those to whom we would offer advice widens. Since extraterrestrial colonies would probably be more different than anything we have experienced before, we should allow them a good bit of latitude. To the extent that we forget this, we may one day hear familiar sentiments from Martian colonists gathered at the air scrubber:

> Those damn wastes of oxygen from Earth! They've never had to endure 3 years cooped up in a space ship or deal with a depressurization emergency, yet they see fit to tell us how to run our business. They treat us like ignorant children and refuse to let us establish families or our own economy. This is our world!

13.5 The Colonial Imperative

If we are confident that terrestrial culture is already well adapted, then what need do we have of further cultural diversity? In that case, why waste resources on colonies? Many commentators thus conclude that we are better off focusing on the many problems humans face on Earth. Yet in the preceding section, we saw how a belief in the universality of terrestrial ideals is problematic from a long-term evolutionary perspective. If we are honest, we must admit that it is far from clear whether we can

[7] I actually share with Stapledon (1948) a conviction in truly universal ethical principles, though unlike most ethicists, I develop these from evolutionary principles (Smith 2014). But this is a sensible position only if the principles in question are couched at a very high level of abstraction, in much the same way that a universal definition of life would have to be abstracted away from biochemical details (Smith 2015c). So while it may be defensible to say that human societies should strive for liberal governments when conditions allow, arguing that they should never engage in strip mining is far too specific.

or should extrapolate from what is best on Earth to what is best in space. But there is also a more subtle form of overconfidence. Perhaps we don't think that we have all the answers *now*, but we believe terrestrial culture will be able to rise to whatever challenge comes along without assistance. There are two possible attitudes behind this overconfidence that bear critical examination.

First, there is what might be called the historical attitude: our terrestrial culture has always managed to adapt in the past, so we know it will continue to do so in future. This is problematic for several reasons. For one, it's not true that *terrestrial* culture has successfully adapted, since terrestrial culture as a whole has never been subject to natural selection before. It's more accurate to say that specific *subsets* of terrestrial culture (e.g., 18th century British, French and American) have successfully adapted in the past. But not all of these cultures were successful—indeed, some cultures have vanished altogether (e.g., the Shakers). Therefore, not only is there no sense in which *terrestrial* culture has defeated all comers, but failure of *individual* terrestrial cultures is sufficiently commonplace in history that broad optimism does not seem warranted.

Even if the hypothesis of historical success could somehow be salvaged, past success is no guarantee of future performance. This is particularly true if we admit that we will eventually face adaptive challenges radically different from what we have encountered before. The dinosaurs ruled the Earth for 135 million years—roughly 650 times longer than modern humans have been in existence. Had they been able to express themselves, they would doubtless have voiced very high levels of confidence in their ability to cope with future challenges—right up to the Chicxulub impact. It is therefore either naked hubris or extremely short-term thinking to assume that human culture as it stands will be able to overcome every threat it may face in the future. The only reason this is not more obvious is that people have a difficult time seeing past their present circumstances to imagine a fundamentally new type of challenge. Such lack of imagination can be very dangerous. Consider those who climb on giant boulders perched on the edges of cliffs, feeling secure despite the evidence of previous rock falls. Since they can't imagine something so large that's been there so long suddenly shifting and falling into the valley below, they are utterly astonished when their weight causes the rock to give way, as it occasionally does.

The second problematic attitude I will call dynamic confidence: people may grant that *present* human culture may have shortcomings, but nevertheless express confidence that, since it evolves so rapidly, it will be able to rise to meet new challenges. The basic problem here is that the people judging the pace and promise of cultural adaptation are from within the culture. It's therefore quite difficult for them to judge the extent to which the change they see is *objectively meaningful*. Consider the dystopian movie *WALL-E*, which features a generation ship that's been journeying through space for 700 years. During this time, the crew have developed a completely hedonistic culture, taking advantage of the many diversions their ship provides to fill their lives with distractions. They cease caring about their mission and gradually lose the skills they will need to survive anywhere outside the ship. From an external perspective, this is clearly maladaptive in the long run as the ship can not sustain itself forever. But a member of the crew, hearing her culture so accused would defend it vigorously by listing the many major changes her society has experienced in her lifetime. Of course, these would all be objectively trivial things (e.g., the evolution of public taste in movies) that are important only subjectively, within the context of the culture. And are not adaptive in any objective fashion—but this would be hard for a member of that culture to discern. Thus, our subjective sense that our culture is evolving rapidly is, at best, extremely weak evidence that it possesses the kind of dynamism needed to meet fundamentally new types of challenges.

To make matters worse, terrestrial culture may be headed toward a period of cultural stagnation in the absence of active measures to foster adaptive diversity. As we develop new technologies for travel and communication, our culture will likely become increasingly homogenized, trending towards a monoculture. Since natural selection requires competition, monocultures are bereft of its guiding influence. Worrying about cultural stagnation may seem an odd claim when it's widely accepted that one of the hallmarks of modern civilization is rapid, ever-present, change. But the problem is not that a monoculture is static, it's that what change does occur is uncorrelated with the adaptive needs of the future.

13.6 Conclusion

We are ultimately just as myopic as any other population at the bottom of a local optimum. The conservatives may be right that we will not encounter challenges we can't meet, at least for a long while. But then again, they may be wrong. If they are, the consequences could be disastrous. In other words, we should be *worried* rather than complacent.

Human culture on Earth is not unlike an alpine ecosystem living in a fragile and unique valley, surrounded by steep mountain ranges on all sides. Historically, the mountains have protected the valley from invasion, but they also limit adaptive change. Things have gone well for the inhabitants of the valley—the environment has always been relatively benign and as a result they have prospered and diversified, gradually filling the local area with a complex network of living beings. From the perspective of the valley dwellers, the future looks bright. The problem, of course, is that things are likely to change sooner or later and, when they do, there is a good chance the valley ecosystem will wither and die. If your job were to insure the long-term survival of these organisms, you would almost certainly undertake an aggressive colonization program, forcing the inhabitants to live outside their traditional, comfortable, borders. Earth must do the same.

The observation that we live in an isolated and fragile environment is often made, particularly by those who have seen Earth from space (Northcott 2007). Ironically, this fact is commonly used as an argument *against* space exploration. The thought is that, since Earth is so fragile, we should direct our efforts to making sure we preserve its essential attributes. Putting the future of humanity in a single basket on the grounds that it is a fragile one is not the wisest long-term strategy, however. What's more, there is clearly an implicit assumption which is highly questionable, namely that all the resources we need to preserve the Earth will be found on Earth.[8]

I have talked about unspecified threats to humanity, partly in order to avoid being bogged down in detail, and partly because we must worry about threats we can't even identify at the moment. But it's worth noting that there are any number of perfectly ordinary (from a cosmic point of view) threats that could wipe out human life on Earth: asteroids, pandemics, war, global warming, etc. (Smith 2008; Bostrum and Cirkovic 2011). Any one of these threats could easily destroy humanity outright, or at the very least knock human civilization back to the stone age, causing untold suffering and death in the process.[9] Of course, humans were not selected for our ability to engage in truly long term planning, so it's psychologically easy for us to ignore threats that may not materialize for thousands of years (Yudkowsky 2008). But we have excellent reasons to think that it's only a matter of time before the universe calls our bluff and we will be forced to adapt like never before. Our failure to see colonization as an imperative in the face of clear, if distant, threats is an excellent example of how our present adaptations do not always prepare us for future challenges.

If we wish to overcome our terrestrial provincialism, as we ultimately must, we need to pursue strategies to foster and maintain a pool of adaptive diversity to help us meet far greater challenges than we have encountered in the past. We simply can't afford the cultural opportunity costs and long-term risk of conservative myopia. Thus, any strategic plan for humanity worthy of the name must include the aggressive establishment of extraterrestrial colonies when this becomes practical.

[8]There are also opportunity costs with clear moral implications we incur by confining ourselves to a closed terrestrial system in which sustainability considerations will eventually force us to impose restrictions on population, standard of living and resource intensive activities such as big science.

[9]It should be noted for those who object to such an anthropocentric point of view that any event with such a devastating effect on humans would also wreck havoc on the terrestrial ecosystem more generally.

References

Adams, E. (2006). Rods from God: Space-launched darts that strike like meteors. *Popular Science, 4.*

Bostrom, N., & Cirkovic, M. M. (2011). *Global catastrophic risks.* Oxford: Oxford University Press.

Boyd, R. (1991). Realism, anti-foundationalism, and the enthusiasm for natural kinds. *Philosophical Studies, 61*(1/2), 127–148.

Carroll, L. (2014) *Through the looking glass.* CreateSpace.

Cavalli-Sforza, L. L., & Feldman, M. W. (1981). *Cultural transmission and evolution.* Princeton, NJ: Princeton University Press.

Cockell, C. (2002). Mars is an awful place to live. *Interdisciplinary Science Reviews, 27*(1), 32–38.

Dyson, F. (2014). *The scientist as Rebel.* New York: New York Review Books.

Efstratiadis, A., et al. (1980). The structure and evolution of the human β-globin gene family. *Cell, 21*(3), 653–668.

Elkins, S., & McKitrick, E. (1954). A meaning for Turner's Frontier. *Political Science Quarterly, 49,* 321–353.

Grant, M. J., Ross, A. S., Button, C. M., Hannah, T. E., & Hoskins, R. (2001). Attitudes and stereotypes about attitudes across the lifespan. *Social Behavior and Personality, 29,* 749–762.

Henningham, J. P. (1996). A 12-item scale of social conservatism. *Personality and Individual Differences, 20,* 517–519.

Irwin, G. (1989). Against, across and down the wind: A case for the systematic exploration of the remote Pacific Islands. *The Journal of the Polynesian Society,* 167–206.

Kuhn, T. S. (2012). *The structure of scientific revolutions.* Chicago: University of Chicago Press.

Landes, D. (1998). *The wealth and poverty of nations.* WW Norton & Company.

Limerick, P. N. (1987). *The legacy of conquest: The unbroken past of the American West.* New York: WW Norton & Company.

Määttä, S. (2005). The European charter for regional or minority languages, French language laws, and national identity. *Language Policy, 4*(2), 167–186.

Maltby, J. (1997). The concurrent validity of a short measure of social conservatism among English students. *Personality and Individual Differences, 23,* 901–903.

Mesoudi, A., & Whiten, A. (2008). The multiple roles of cultural transmission experiments in understanding human cultural evolution. *Philosophical Transactions of the Royal Society B: Biological Sciences, 363*(1509), 3489–3501.

Mill, J. S. (1977). *Essay on politics and society.* Toronto: University of Toronto Press.

Northcott, M. S. (2007). From environmental utopianism to parochial ecology: Communities of place and the politics of sustainability. *Journal for the Study of Religion, Nature and Culture, 5* (1), 71–85.

Pierson, G. W. (1942). The Frontier and American Institutions: A criticism of the turner theory. *New England Quarterly, 15,* 224–255.

Riley, G. (1993). Frederick Jackson Turner overlooked the ladies. *Journal of the Early Republic, 13,* 216–230.

Rollin, B. E. (1995). *The Frankenstein syndrome: Ethical and social issues in the genetic engineering of animals.* Cambridge: Cambridge University Press.

Rosenberg, A. (2007). *Philosophy of biology: A contemporary introduction.* London: Routledge.

Rosenberg, E. (2011). *Spreading the American dream: American economic and cultural expansion, 1890–1945.* New York: Macmillan.

Schurman, R., & Munro, W. (2006). Ideas, thinkers, and social networks: The process of grievance construction in the anti-genetic engineering movement. *Theory and Society, 35*(1), 1–38.

Sekiguchi, K., & Shuster, C. N., Jr. (2009). Limits on the global distribution of horseshoe crabs (Limulacea): Lessons learned from two lifetimes of observations: Asia and America.

In J. T. Tanacredi, M. L. Botton & D. Smith (Eds.), *Biology and conservation of Horseshoe crabs* (pp. 5–24). New York: Springer.
Smith, K. C. (2008). The terrestrial lifeboat project: An international undertaking to safeguard humanity. In Y. An (Ed.), *A new perspective on applied ethics* (pp. 224–238). Renmin Press (in Chinese—English translation available on author's researchgate site).
Smith, K. C. (2014). Manifest complexity: A foundational ethics for astrobiology? *Space Policy, 30*(4), 209–214.
Smith, K. C. (2015a). Notes from *The ethics of human activity in outer space* workshop, Hatfield, UK, June 2015.
Smith, K. C. (2015b). Notes from *Extraterrestrial liberty III: Dissent, revolution and liberty in space* workshop, London, UK, June 2015.
Smith, K. C. (2015c). Life is hard: Countering definitional pessimism concerning the definition of life. Manuscript submitted to *International Journal of Astrobiology*, 9/15.
Stapledon, O. (1948). Interplanetary Man? *Journal of the British Interplanetary Society, 7*, 213–233.
Turner, F. J. (2013). *The Frontier in American history*. CreateSpace.
Van Valen, L. (1973). A new evolutionary law. *Evolutionary Theory, 1*, 1–30.
Via, S. (2001). Sympatric speciation in animals: the ugly duckling grows up. *Trends in Ecology and Evolution, 16*(7), 381–390.
Wade, R. C. (1996). *The urban frontier: The rise of western cities, 1790–1830*. Urbana: University of Illinois Press.
Wilson, M., & Daly, M. (1985). Competitiveness, risk taking, and violence: The young male syndrome. *Ethology and Sociobiology, 6*(1), 59–73.
Yudkowsky, E. (2008). Cognitive biases potentially affecting judgment of global risks. *Global Catastrophic Risks, 1*, 86.

Chapter 14
Overlords, Vassals, Serfs? How Space Colonies, the Future of the Space Economy and Feudalism Are Connected

Andrew Kennedy

Abstract The purpose of this paper is to examine what economic, social and biological seeds we need to sow to create successful space colonies better protected from dissent and revolution. It considers that while the space development timetable is continuing, the Earth economy in its current form will damage the biosphere before space industrialisation can be established, and without a healthy consumer economy on Earth, space colonisation is unlikely to occur. Using themes drawn from earlier feudal structures and considering the birth of capitalism and the nature of growth, this paper examines three long-standing assumptions about the drivers of the space economy. The first driver is that the space industrialisation is a necessary stimulus to the Earth economy. The second is the obligation of humanity to physically diversify to save itself from extinction, and the third is humans' innate exploratory nature, which must be given full expression. This paper will show that none of these drivers will be successful in altering the current economic realities of the space economy and that in particular, the third driver also misrepresents how humans explore. It will show that the Earth degradation timetable and the space economy development timetable do not match, and that the Earth's biosphere is likely to become irreparably damaged long before the space economy can support it or become self sustaining. This paper considers that solutions to the problems of biosphere degradation and sustainable development on Earth will be the same solutions to those of colonising space, and will describe a two-part implementation of a scheme to provide a secure foundation for successful space colonies. Firstly, by implementing historical features of human societies to enable natural decision-making procedures to develop among diverse groups dedicated to space development and secondly to slowly separate the space economy from the world economy of fiat currency and standard capitalist investments vehicles, which will direct the evolution of space colonies along a path compatible with both a protected biosphere on Earth and long lasting settlements away from it.

A. Kennedy (✉)
Chronolith Foundation, Calle Goles 48, bajo, 41002 Seville, Spain
e-mail: ankank@inicia.es

C.S. Cockell (ed.), *Dissent, Revolution and Liberty Beyond Earth*,
Space and Society, DOI 10.1007/978-3-319-29349-3_14

Keywords Space economy · Development timetable · Capitalism · Feudalism · Biosphere · Timetable · Non-state actors · Polycentric decision-making · Crypto-currency

14.1 Introduction

When it comes to the future establishment of space colonies, the kinds of societies that will inhabit them, the rights and liberties those inhabitants will have and the mechanisms they will use to express dissent and make and change decisions are questions that appear to bear directly on the success or failure of such colonies (Cockell 2013). This paper will examine some economic and social foundations to the colonising of space and to consider the appropriate seeds we need to sow now for the colonisation process to begin on a sure footing and long before humans actually begin to occupy locations in space.

The often-expressed view of future visionaries is that the space economy should be driven by the need to save humanity from extinction. For example, Stephen Hawking: "mankind must colonise space or die out." (Hawking 2010); or Elon Musk, "…I think there is a strong humanitarian argument for making life multi-planetary…in order to safeguard the existence of humanity in the event that something catastrophic were to happen…." (Musk 2004).

Others point out that the technological developments in the pipeline will drive the space economy. Industries founded on autonomous robots, programmable shape-shifting materials, nano devices like fluid computers, movement amplifier suits for workers in space, artificial organs and genetically engineered plants and animals for life in space, advanced health care and organ regeneration and more, are innovations that will make space habitable, even though the most significant of these are projected to be in use only well into the second half of the century. This paper will consider however, that the current economics of capital and labour are more determining of humans' ultimate progress in space than any of these drivers.

It seems that in many of the discussions about settlements in space, the Earth economy is expected to fund space industrialisation naturally (Adam Smith's 'invisible hand'), but the economic realities do not seem to support this. While there are economic benefits for governments to invest in space activities, generally considered to be an investment multiplier returning value by way of new products, advanced cutting edge research and employment for every $1 spent, they do not yet obey market forces as a whole, and the route by which general benefits from investment in space activities accrue to the Earth economy are often assumed but not clearly drawn and are still controversial.

An optimistic approach to space settlements found in the NASA study (NASA 1977), the result of design work initiated by Gerard O'Neil in 1975, was to put

10,000 people into an L5 habitat in 20 years from go-ahead, and attempted to calculate the cost of this goal. Yet the matter of who paid for this and who received the income from the activities was not described. The Earth market for electricity and metallic resources is assumed to monetise the new supply from space at ever rising values unencumbered by rising costs from, say, biosphere degradation and/or feeble consumer growth and excessive debt. The report from the US National Commission On Space, ten years later, (US Commission on Space 1986) said nothing about the Earth economy and the future risks to its biosphere from human activities as it assessed space settlement strategies. It considered the space economy only in terms of the potential for space science. Even a discussion paper published in 2015 (Ferguson et al. 2015) mentioned nothing about how the markets might actually produce the anticipated habitat in space, housing 300 people, by AD2050. NASA itself in its 2015 study of manned missions to Mars did not analyse where the investment funds were going to come from nor what returns for investors might be created to stimulate such investment and said only that it would implement them "…with budgets commensurate with economic growth." (NASA 2015a).

A self-sustaining presence of humans in space must begin with the Earth economy and a means by which it can pass on to the inhabitants of space at least some of its powers of consumption. The question of whether the Earth can maintain its own consumption sufficiently to allow the space economy to mature has not been properly addressed. Economists recognise the threat of the Hoyle Trap (1964), a certain point in development where if growth falls back, it can never regain the same level because the resources required will have been used up or extracting the remaining ones require a level of advancement that can no longer be reached (Kennedy 2013). Hoyle simplified the dilemma, however: either we push for growth or we fail. He did not seem to consider that we might push for growth and still fail because of the damage to the Earth that growth produces.

No one doubts that we stand at risk of actual extinction from man-made threats ranging from political failures leading to nuclear or biochemical warfare to technological and bioengineering errors or from unpredictable natural events like volcanic eruptions or meteor bombardment. But the extinction path plotted from economic mismanagement, biosphere damage and in particular from the loss of resources, though complex, is easier to establish. Some economists are still optimistic about resource depletion and believe that the market will raise prices such that there will always be resources remaining. While this may be true in a rising market available resources are not the only factor to limitations to growth. During a deflationary episode, no amount of price adjustment can create a demand where there is none and economies can still fail when the cost of resource extraction exceeds funds available for investment, whether or not resource limits have been reached. Further, some resources, like phosphorous, or pollinating insects, or organic soil itself are more crucial to economic growth than others and a few strategic depletions may cause irreparable damage to the growth rate even while other resources remain. As a simple example of how at risk humans are to the side effects of growth can be seen from a calculation from the WorldWatch organisation which shows that the planet has available 1.9 hectares of biologically productive land per person to supply food and other biologically active resources and absorb wastes, yet the average person on Earth already uses 2.3 ha, ranging from the 9.7 ha used by the average American to the 0.47 ha used by the average Mozambican (Worldwatch 2013).

Hoyle's observation, that we have one chance to make it as an extraterrestrial civilisation, seems like an imperative to maintain our economic momentum at whatever cost. Yet, there is serious concern about the negative impact on human survivability from growth regardless of any economic potential that exists from the opening up of space to human activity. Resource depletion is still only part of the story, and other effects of growth such as the damage to the capacity of humans to survive as a whole from waste, unpredicted disease and climate change also adversely affect the global economy. In this paper I shall show that the timescale of space industrialisation is unlikely to get humans well established in space before the Hoyle trap begins to close. I shall show that our ability to found space colonies is becoming increasingly remote as the relationship between capital and labour changes, making the space economy less attractive to capital for investment and inhibiting capitalist investment cycles from closing the funding gap between government spending and the private sector. It considers, in particular how both the feudal origins of capitalist growth and the natural exponential productivity of the biosphere seeded an economic system that cannot effectively produce the colonisation of space without significant changes to it. This analysis does not consider whether or not humans will meet the technological challenges of space exploration eventually, but only that the timetable for development will be far too slow and that, without first maintaining and protecting the health of the Earth's biosphere and its biological productivity, the Earth economy will not be able to support sufficient space industrialisation to save itself or supply a safe haven for humanity. In conclusion, some solutions to this crisis will be proposed and a strategic plan for creating space colonies free of damaging trends inherited from the current competitive economic model will be derived from them.

14.2 Part 2: Space Development Timetable

14.2.1 The Next 100 Years

By looking at what the space industry is already planning, we can begin to see how seriously the development of the space economy is lagging behind events in the Earth economy, where lack of political will, of the implementation of technology and of economic creativity leaves a high probability that it is unsustainable in the long term and unable to support the necessary elements of that space economy. By plotting the timing of the most likely space economy 'events' against the timing of highly probable events in the progress of the Earth economy we can observe the situation more clearly.

While precise estimates of the timing of resource depletions are difficult to establish through future economic cycles, we can take a single mineral resource, phosphorous, and allow it to give us a bound to our developmental timetable. Phosphorous is reasonably abundant on Earth, though in the Solar System as a whole it is in relatively short supply. Phosphate is required by every living thing and its use in agriculture is perhaps the most critical for the future of humans. Some think that the world's known phosphate deposits will be exhausted by the end of the century (Vaccari 2009). Other scientists are confident that reserves will last a few hundred years (IFDC 2010). Current reserves, however, are reducing in quality and many are contaminated with metals like cadmium and uranium which require more energy to clean, and there is also the fossil energy use that comes from transporting over 170 million tons of fertiliser around the world. Yet there is no coordinate plan to conserve it. It is wasted all over the world and it pollutes agricultural run off that contributes to aquatic dead zones, both in fresh water supplies and in seas, that are growing by 10 % a decade (Greenpeace Research Laboratories 2008). In the absence of efficient re-cycling, we need increasing amounts of phosphorus fertilizer to maintain food production for all other living creatures as well as humans. 100 years from now will give us the temporal range on which to plot the points of our space economy. (Other resource limits will be reached much earlier.)

14.2.2 The Space Economy Timetable in Broad Strokes

Let us consider the principle likely income generating events of the space development timetable already in the planning stages and with a greater than 50 % chance of occurring within a 5 years spread around the projected date and compare their occurrence with what is projected to occur in the Earth's biosphere during the same period (Fig. 14.1). The timetable does not include various science missions or non-manned preparatory flights. There will be more probes to Mars and the other planets but none of these will impact significantly on the colonising ventures in the Solar system for a long time to come.

14.2.2.1 AD2015—Satellite Launches—on-Going

> In LEO the single biggest source of private benefit is the Earth observation sector, with an estimated total value in 2011 of $2.24 billion. However, this revenue stems largely from civil government and military customers, and governments are expected to fund development of most Earth observation satellites during the next ten years. Aside from Earth

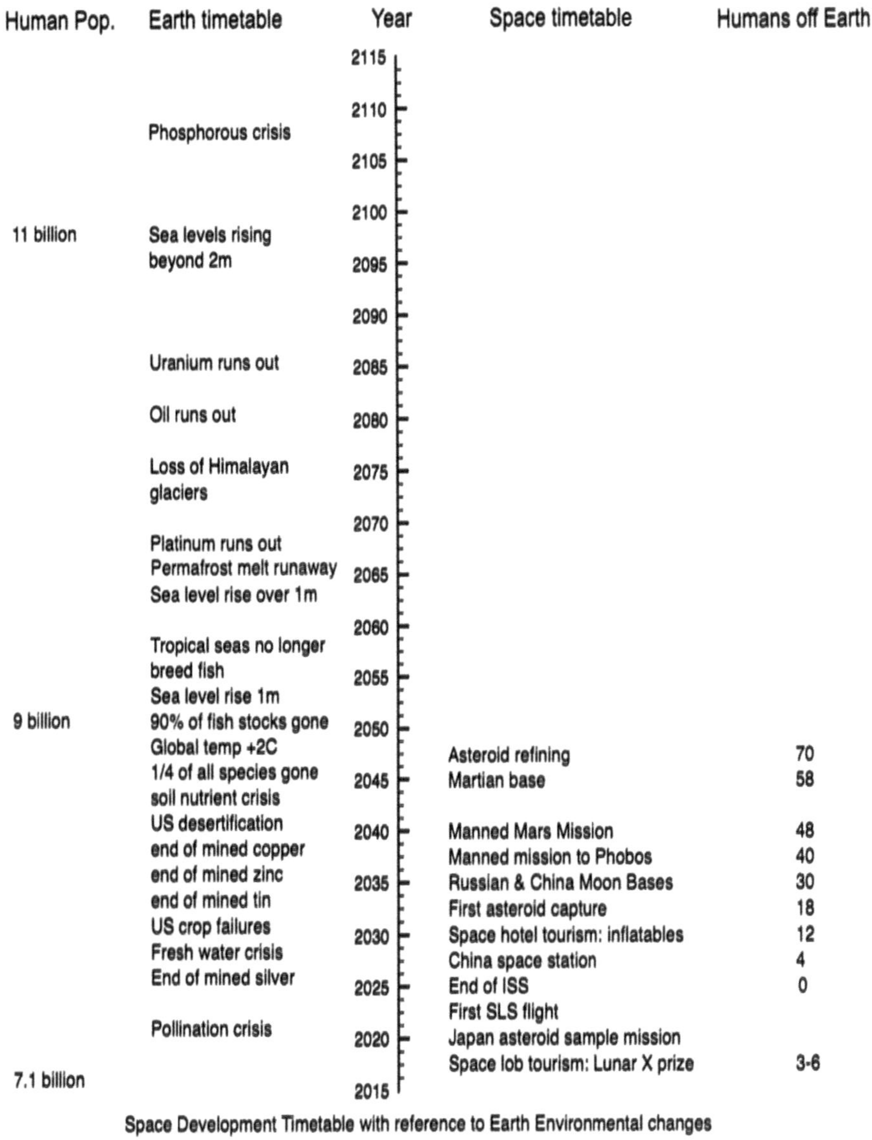

Fig. 14.1 Activity in space compared with significant Earth environment changes over the next 100 years

observation, the only other meaningful revenue from satellites in LEO comes from three communications satellite constellations: Globalstar (46 satellites), ORBCOMM (26 satellites), and Iridium (71 satellites). Collectively, these firms have annual earnings on the order of $100 billion. Data from the insurance industry supports this as well: only 24 satellites in LEO are currently carrying commercial insurance, for a total insured value of about $1.4 billion, out of a total satellite insurance market of $20 billion. (Weeden 2012)

In 2015, there are almost 1300 working satellites around Earth in various stages of obsolescence, about 900 of these split almost 50–50 between LEO and geosynchronous orbit. The Satellite Industry Association give total revenues for 2015 at $203 billion, slightly down on 2014. Since 2009, the market for new communications satellites has been growing at almost 40 % fuelled by technological developments such as the spot beam system that have opened up direct satellite to home possibilities and requiring the replacement of ageing communications satellites. In spite of the trend towards cheaper 'cubesat' manufacture, the costs of manufacturing, launching and using a sophisticated communication satellite is still large. They require sophisticated equipment like transponders and ground stations are expensive to build and maintain. Constellation satellite proposals to bring the internet to the rural half of the world that does not have it are still uncertain financial propositions. The faster the connection the more it costs since users get a bundled service and will use more through video downloads etc. while income does not rise. This problem is less pronounced for 2G cellular networks, where the revenues for voice and SMS services, can more easily justify rural telephony, as bandwidth usage is closely linked to revenue generating and profitable voice and SMS services. This problem worsens as the band width rises and population density in rural areas falls, inhibiting the introduction of even faster 5G networks. Technical problems of latency have not yet been solved. While Space X is continuing its plans for a constellation network, Facebook and Google have scrapped theirs. The key technologies for constellation systems are still not in place. For example, the OneWeb enterprise raised $500 million from a consortium of interested parties and placed launch orders with Arianspace, and with Virgin Galactic for its yet untried LaunchOne system of launching small rockets from its White Knight aircraft. The money was declared for 'further key technologies' for a system that is still not yet ready. Facebook is, however, pursuing its two-tier "Internet.org" scheme to provide a preliminary internet access to rural populations in Africa who can use off-the-shelf equipment to connect to the network, and plans to launch the first satellite in 2016, while Google has plans to create a global broad-band network of high altitude balloons; the real intentions of which will become clearer in Part 3.

It is interesting to consider Australia's $1.8 billion attempt to bring broadband to just an estimated 200,000 people in remote rural areas. The first satellite in the two-satellite plan, Sky Muster, was launched 2nd October 2015. It is one of the biggest and most modern communications satellites ever launched yet it will manage to deliver to service providers (not end users) download speeds of only 25 mbps. The network requires building 10 ground stations while its capacity remains forever fixed.

Investment in satellites is still precarious. For example, the bankrupcy of Autralian satellite operator NewSat, in 2015, caused losses in the U.S. Export-Import Bank and France's Coface, the world's two most-active export agencies (ECAs) in funding satellite projects. As a result satellite production has slowed. The third launch failure in the last three years of the Russian Proton system has led the commercial enterprise Immarsat to delay its Global Xpress service and drastically revise its earnings for 2015.

The reality of resource observation satellites as a conservation tool is somewhat different. Observation of resources say, forests, may reveal illegal logging but it also discovers the most productive areas and puts them at risk of intensive exploitation. To pay back investment, the advanced knowledge of states of rainfall and harvests is useful for governments and exporters to operate in the commodities futures markets which stimulates the hedge fund market. In this sense these satellites are tools of speculation as much as that of resource management and can lead to resource exhaustion rather than conservation. Weather prediction is an important source of income for satellite operators allowing savings in transport costs and provision of services. Although weather related savings for insurance, and information on short timescale loads for energy producers are real gains, surprise events like earthquakes and tsunamis cannot be predicted ahead of time, nor can precise values be put on predicted damage from extreme weather and which limits their value.

14.2.2.2 AD 2017–2020 Space Tourism—Suborbital Lobs, AD 2016–2020

Professor Impey seems to think that there is between $50 and 100 billion income available from space tourism by the time we get to orbital space journeys (Impey 2015). However this figure comes from a poll of high school students (possible future clients) asking what they would pay for a trip. There are two principal enterprises with published plans to offer suborbital lobs to the general public, Virgin Galactic and XCOR, while Blue Origin is developing a manned suborbital capsule but with no concrete timetable to when it may fly with passengers.

Virgin Galactic

Intending to fly 5/6 tourists to 110 km, just over the Karman line. The immediate aftermath of the Virgin Galactic tragic accident and loss of their spaceship SS1 in 2014 was that 24 passengers out of about 700 'future astronauts' who paid $250,000 each for the chance to go into space demanded their money back.

The company (by the end of 2014) has 700 people committed to fly, in contrast to the 65,000 people who had apparently applied for tickets by 2013. The first 100 paid $200,000, and the last 600 paid $250,000. This produces a mere $170 million income. Compare this to SS2 development costs, estimated at over $400 million to 2012. The loss of Spaceship 2 in 2014, however, and a change of motor and construction of new craft have placed an unknown burden of cost on the enterprise. It had expected to begin flights in 2011, but the 2014 tragedy is likely to push back Virgin Galactic's maiden commercial flight by at least another several years, if not indefinitely. That will raise serious funding issues since the business is already thought to have exhausted the initial $200 million from Virgin and $280 million capital investment for a 38 % stake provided by equity partner Aabar of Abu Dhabi

(Whitesides 2010), plus the partial investment in the Spaceport by New Mexico State, as well as advanced ticket sales, and is now being funded entirely from within Virgin Group. The original designer (Burt Rutan) of the flight system has left the project and his company, Scaled Composites, that originally constructed the craft, no longer makes them and has been sold to Grumman Corporation who is planning a larger version to air-launch satellites in competition with the Virgin Galactic LauncherOne scheme.

XCOR

XCOR is a private company so no funding figures are available. It is planning to fly their Lynx craft in suborbital flights to 330,000 ft (100 km), the Karman boundary to space. They have apparently 300 signed up to fly at a cost of $100,000 per ticket, although they raised the price in 2015 to $150,000. The small 9 m craft takes only a single passenger but is unique in that it lands and takes off under its own power and can turn around in 2 h with minimal maintenance. They plan 4 trips a day.

14.2.2.3 AD 2020–2040 Space Tourist Trips to Space Stations or Orbiting 'hotel'

There are a number of craft being developed to take humans into orbit and perhaps for stays in the ISS, namely, SpaceX and its 7-seater Dragon capsule; Sierra Nevada and Dream Chaser; Boeing's CST-100 for max 7 astronauts, Orbital Sciences' Cygnus (no NASA funding) and the Orion capsule designed to sit atop NASA's SLS. No figures have been released by any company about the costs and likely price of such orbital trips. It is assumed only that these trips are a logical evolution of the space tourist trend. But this proposition is not simple. There are only two space stations that we expect to be available over the next twenty years, the ISS and the Chinese Tiangong system, which will be smaller.

Currently, training visitors to the ISS takes at least 6 months. The issues of safety and capability are significant. One expects that visitors to the ISS should be able to read Russian and to have familiarity with the escape craft and how to fly them. But the ISS astronauts will also have to be familiar with the tourist craft as well so that they will be able to use it in an emergency. This must reduce the launch frequency, so the actual numbers of tourists reaching the ISS will be limited for some time. In the past tourists to the ISS went up as a single member of a government funded crew, their costs were reduced in proportion. For a full crew of tourists without government funding, the launch and retrieval costs will be very large. There are no figures available to analyse quite how this venture will be profitable. Even billionaire philanthropists are likely to baulk at participating more than once and there are limited numbers of those.

14.2.2.4 AD 2015–2030 Space Stations in LEO

The question of whether the ISS will survive to welcome these tourists is a genuine concern. The Russians are backing the ISS TO 2024 even though they have declared it is getting old, and there's no more science to be done. For a space station of mass around 460,000 kg, habitable volume of 388 cubic meters, it still requires launches of servicing modules supplying several tons every month. The ISS is a space station of two halves: a Russian series of modules and a US series of modules, and they are not very compatible. Russian cosmonauts tend to stay in their 'half' and where a rescue Soyuz is berthed permanently. Problems of compatibility with the many national programs have not yet been solved. Micro cracks are appearing in the structure and more and more small systems break downs are occurring—like false alarms. Apart from work on the health of astronauts and the effects of microgravity on longevity, the ISS's major role is as an educator. In terms of science, NASA has placed US astronaut, Kelly in the ISS for a year early 2015 to study physiological differences between him and his twin remaining on Earth, and there is a proposal to attached a centrifuge module to try out sleeping in artificial gravity, though loss of station keeping due to gyroscopic effects and the difficulty of manufacturing a rotating collar seal that will enable an independent module to rotate while attached to a space station make this proposition unlikely. Bigelow systems wants to put up more inflatable modules (having already tested two). But there is not much science on the horizon for the ISS.

The Chinese intend to place a somewhat smaller space station, the Tiangong project, in orbit by 2022, but with almost half the ISS volume it is projected to support only 3 astronauts with short term capacity for 6. Other orbital tourism projects have come to nothing. So the maximum number of tourists accommodated by current projects may be just 3 in the ISS and perhaps 1 or 2 in the Chinese station (Sample and Branigan 2011).

14.2.2.5 AD 2018–2020 Asteroid Sample and Return

The Japanese space agency JAXA launched Hayabusa 2 spacecraft last year, intending to arrive at asteroid arrive at the asteroid Ryugu in 2018, collect samples and return to Earth in 2020.

14.2.2.6 AD 2017–2040 Moon Mining Facilities

Google's Lunar X Prize, an offering of $20 million for landing a lunar rover, and other prizes up to $30 m in total. Google has just extended the deadline a second time to 2017. Currently (2015) there are 16 teams in the race. Astrobotics Technology, founded in 2008, has booked a place on SpaceX's Falcon 9 rocket for a lunar rover called *Polaris*. Originally it was to launch in October 2015, but this has been set back to 2016. *Polaris* will embark on a one-way mission, carrying NASA instruments to hunt for water, oxygen and methane at the Moon's north pole. *Polaris* benefited from $3.6 m of NASA contracts. But the plan has mutated somewhat and now places on their mission are being sold to others. Moon Express has contracted with a New Zealand rocket maker, Rocket Lab, to launch robotic missions planning eventually to return material to Earth. Lunar Mission One has raised $1 million through crowd funding to build, launch and land a probe at a Lunar pole and drill down below the surface to study its composition and leave personal digital records behind. This sum falls well below even the cost of a launch. It seems improbable that crowd funding could approach the billions needed to perform the full specification mission, even if a launch is shared with others. Moon ores are composites not typically found on Earth requiring investment in new techniques. For example, low gravity is a problem for the convection of molten masses typically required with Earth-based refining. New approaches have been proposed, but all need vast quantities of energy and supplies of hydrogen and carbon dioxide, requiring at least a two stage treatment process of the regolith to first obtain the reducing gasses needed and then use them in the refining process.

NASA has no concrete plans to return to the Moon or develop a Moon base, although studies have been done that suggest a commercial/NASA mix of funds could develop one for creating propellant stocks in about 10–12 years after first landing on the Moon again (Miller et al. 2015). Since their heavy lifting rocket version required for a Lunar Mission is unlikely to be ready before 2023 (first flight of the Orion/SLS combination has been put back 2 years from 2021 to 2023), any date before 2035 for a US Moon base is extremely optimistic. In its own documents NASA is planning only one flight per year of its new SLS rocket throughout the 2020s.

The question of crew safety during moon missions has not been addressed. Up to now manned space missions include a means of returning to Earth in an emergency for all the crew. Since the Soyuz capsule takes only three people, when there is a full complement of 6 astronauts in the ISS, a second Soyuz remains docked so that all could escape in an emergency. This principle will have to be abandoned in any Moon base scenario since the costs of providing escape rockets for all the base personnel will be prohibitive. In any event, where does the Moon base crew go in the event of a critical emergency? Staying in Moon orbit is not a permanent solution until a station in Lunar space has been built, so they would have to return to Earth which means a craft would need to be ready for them to do this either in Moon orbit or in Earth orbit, and this provision will be way beyond the mission parameters of any country for some time to come. The impact of this difficulty is clear. If human colonists are simply to be abandoned in emergencies on the Moon or on Mars this implies a huge loss of investment return, further inhibiting the deployment of funds for such missions.

14.2.2.7 AD 2020 Space Hotel

There are expectations that an LEO space 'hotel' will be in place by 2050, but no concrete plans have been laid. Questions of escape capsules and the enormous re-supply cost problem have not been solved. Consider the ISS. Originally planned for 2004 completion, the first module was launched in 1998 on a Proton system. It took 27 Shuttle launches, 7 Soyuz and Proton launches and some half dozen others to date to assemble the station as it is now. (Other shuttle launches did not bring any station hardware). At around $1.5 bn total of investment averaged per Shuttle launch we are looking at say $50 bn round figures in 2008 dollars. Development and running costs over ten years comes to $100 bn. Bigelow Aerospace has a 330 cubic metre inflatable module B330 with which it plans to construct space stations with occupancy for 6 and a 20 year lifespan. With a NASA grant, they will test a smaller activity module to be attached to the ISS in 2015 but this date is uncertain. A space hotel will need be a more complicated structure than the ISS if it is to be attractive enough to justify the high cost of flights to it and the high costs of maintenance.

14.2.2.8 AD 2021–2040 Asteroid Capture and Mining

NASA plans to bring back 2 kg of asteroid material in 2021 using robotic systems. Planetary Resources plans to launch several small satellites to survey the near Earth asteroid population. So far they have launched Arkyd study probe which is currently at the ISS for launching into LEO some time later in 2015, and it plans exploratory missions in the 2020s but has no concrete missions to extract value from them. Even though on Earth readily accessible supplies of even ordinary metals like silver, lead and platinum are already running down, the actual cost of space resourced metal is likely to put it well beyond the reach and need of the Earth economy. If a resource becomes too expensive then devising alternatives or ways of doing without it are going to be cheaper and probably quicker to develop. The case for off-Earth mining makes sense for off-Earth construction only, and this will be investment money lost to Earth. The return to Earth of metal payloads is unlikely to make any economic sense. Even a re-usable vehicle is unable to carry much by way of payload through re-entry since heat generated during it is proportional to the mass.

14.2.2.9 AD 2035 Mars Short Stay

The problems with going to Mars are well known. With current and foreseeable technology the most dependable and economic minimum duration stay Mars trip gives 3–4 months on the surface using Hofmann transfer orbits. A return trip to Mars would take about 21 months: 9 months to get there, 3 months on the surface, and 9 months return.

Mars crews will need to bring enough food, water, clothes, and medical supplies in addition to all the scientific instruments, and all the fuel they will need for the return. It is not known even how fuel stocks can be maintained for such long periods of time. Unless cryogenic capabilities are among the first facilities constructed on the surface of Mars, then solid fuel rockets will have to be brought from Earth and landed on Mars. For a nine month voyage the crew needs a lot of radiation shielding. Water, and even cement, make good shielding but they are very heavy and eventually their utility falls as they become transmitters of radiation themselves. It is estimated that for a crew of six, you would need around 3 million pounds of supplies. At a launch of say 50,000 pounds of payload, that is 60 launches alone (next generation rockets will lift 100,000 lbs but will fly less often). The spacecraft itself has to be launched in bits and then assembled, with separate launches for the crew. Assuming some form of craft along the lines of the minimum volume of the ISS + surface craft and living modules, we can expect there to be at least twice the ISS all up weight of 465 tons or say another 1.9 million lbs given the added weight of a nuclear power plant since the solar radiance at Mars is about 250 times less than at LEO and solar panels will be insufficient. The ISS required 40 launches for its assembly and fitting out. So it seems that between 100 and 140 minimum launches depending on the heavy lift capabilities of the SLS (which will not be available until at least 2023, and NASA's own projections appear to have only one launch a year), to get a craft ready for the Mars trip. The Space Shuttle did only 135 launches over its entire 30 year life span. Reusability of rockets is intended to improve upon the economics and turn around time of this output but the various technologies are yet to be proven in the field.

14.2.2.10 AD 2040 Microwave Energy Stations

Plans have been sketched out to manufacture orbiting facilities that collect solar power convert it into microwaves and transmit a narrow beam 24 h a day to the Earth's surface for conversion into electricity to boost the Earth economy manyfold with 'free' energy. None of the technical difficulties associated with this idea have been solved. The lack of stability of focus of the microwave beam is a known problem without a solution. Interference with other satellites, especially with the vast numbers circulating all over orbital space required by constellation systems will make these plans almost impossible to put into action. The square kilometres of solar panels required, or vast mirrors for turbine generators all require maintenance, requiring launches and human workers, while on Earth maintenance of receiving stations makes this 'free' energy anything but free.

14.2.2.11 AD 2040 Mars Long Stay

No concrete plans have yet been produced, except NASA has published a plan to use Phobos as a staging post for Mars exploration in preference to a Mars surface station (NASA 2015b), but surface missions with human explorers designed to stay months or years on the surface have not been outlined in any detail. The Dutch Mars One project originally planned to add 4 people every two years after 2030 has not provided details of how the project will be fully funded but the CEO, Bas Lansdorp, claims that its $6 million budget and income from investments will be sufficient since return trips are not required. Already the planned missions have been delayed by 2 years and further delays seem likely. In the long run staging points on the way to Mars and the outer planets will be required, supplied initially by a Moon industrial base, but this will not come about until the second half of this century.

14.2.3 Space Investment Economic Multiplier

It is generally considered that there is an economic multiplier effect from each dollar spent on space activities which is certainly one of the reasons why there is a growing number of countries entering the space economy. Yet the question must be asked why, if this is true, investment from the private sector is not larger than it currently is, and why the space development timetable has a short and uncertain horizon. We shall examine the role capital economics and its relationship to labour plays in creating this uncertainty, but it is also instructive to consider the validity of the multiplier effect in understanding why investment in space activities has still not been fully embraced by the markets.

The $7–14 of economic gain for every dollar spent (Gurtuna 2013), commonly quoted in the press, is controversial. In earlier years, many supporters of the space program placed great stock in the benefits of technological spinoff from the space effort for the American economy. Later estimates claimed $7 in return from every $1 spent (Lyttle 1991). While a Chase Associate report suggested a $23 billion gain for $1 billion spent by 1986, they also said that $21 billion of that could be attributed to other factors and that overall it reported that (the multiplier) was,

> … not crucial to deciding whether more or less money should be spent on NASA R&D, because similar effects could be obtained by other forms of government spending – such as defence procurement or energy R&D. Tax cuts are, of course, a comparable alternative… (Chase Econometric Associates 1977).

More recently, the European Space Agency has quoted a familiar multiple on its public web site,

> The economic impact of space spending is boosted by a multiplier effect: every €1 invested in space returns an average €6 to the wider economy. So space contributes to growth, employment and competitiveness across many economic sectors, while also being largely immune from outsourcing: (ESA 2015).

However the OECD has leaner figures for its multiplier,

> In the United Kingdom, the space industry's value-added multiplier has been estimated to be 1.91. Finally, the most recent Federal Aviation Administration (FAA) study on the economic impacts of the US commercial space activities has also shown a rather stable multiplier ratio since 2002. In 2009, for every dollar spent in commercial space transportation industry, USD 4.9 resulted in indirect and induced economic impact. (OECD 2011)

While NASA currently has a spin-off web site (NASA 2015a) which furthers these kinds of figures and where one can read about the specific benefits to the economy spun off from its activities, the spin-off arguments in favour of this investment may be flawed since it is hard to separate general economic growth and innovation from NASA-created growth. Such growth depends on markets rather than pure technology solutions. A closer examination of the spinoff record would provide little comfort for space advocates. One German analysis of space spinoffs concluded that,

> The overall conclusion to be drawn from this is that the spin-off rate is very low in highly specialized space projects – a conclusion which coincides with the finds of other investigations. The concept of a decisive spin-off in the narrow, real sense of the term cannot therefore be validated on the result of these findings… many standard examples of spin-offs may be traced back to the first R&D boom in the Sixties… only in the rarest of cases do the spin-offs prove to be identifiable as classic cases in which the source can be associated exclusively with space technology and the diffusion be associated with a sector unrelated to space technology. In the majority of cases, both source and diffusion can be associated with multiple purposes both within and outside space technology. (Schmoch et al. 1991)

The most comprehensive review of the impact of NASA technology benefits to the commercial sector was conducted for NASA by the Chapman Research Group in 1989. This study evaluated the benefits derived from technologies identified in the annual NASA report *Spinoff* during the period from 1978 through 1986 (the deployment of Skylab and introduction of the Shuttle era), and quantified the benefits from the total NASA investment of $55 billion as being $5 billion in true spinoffs (that is, a product or process or company that would not have existed without NASA) suggesting a very poor 1:10 payoff and about 100 times worse than the commercial economy as a whole (Chapman Research Group 1989). Even though these calculations are for past performance and that NASA has changed its spending and management model, this broadly agrees with our attempt to calculate a deployment of capital per employee ratio which we will discuss in the next section of this paper.

The economy of space activities is likely to be dominated by government funds for some time to come with an impact on the speed of advancement of the space economy. Certainly, space has become a worthwhile investment for governments to help their national social, economic and technological development but also to develop national defence and security programs where there is no dividend to pay to investors, and the number of space programs is growing. 26 countries reported a space program in 2001, 42 in 2006, and 57 countries own and operate at least one satellite in 2012. As a result, governments around the world seed advanced research projects where there appear to be fundamental science pay offs, like the Hubble or Rosetta, or things they can immediately sell like satellites or launch capability or to possess an independent capability in communications. Even so, current government budgets of the principle actors in the space economy[1] represent a mere 0.043 % ($46.71 bn) of world GDP of $107 trillion in 2015. Even supposing a stimulus multiplier effect of 1:8 is in operation for all national programs, this government

[1]Excluding military and secret government missions, space programs in billions of dollars: NASA, 18.0; Russia, 8.6; China, 12.0; ESA, 5.0; India, 1.3; Japan, 1.5; UK, 0.3. (CIA World Fact Book 2015).

funding would add only 0.35 % to world GDP. Even if the proportion of private to government investment changes radically, the impact of technology on the timing of the events of the next 35 years of space activities will be small.

14.2.4 Earth Environmental Timetable

What will happen to the Earth's biosphere and global economy during the next 100 years, however, is much better understood. By the year 2050, Earth will be struggling with a population of 9 billion or more (UNDEA 2015), with an average global temperature rise of perhaps 2 °C along with a sea level rise of as much as a metre (IPCC 2014), and while extractable stocks of metals available in the long run cannot be easily calculated (Graedel et al. 2011), in the short run, within 35 years, there will be depleted or exhausted stocks of readily extractable tin, copper, zinc, and gold, few fish in tropical waters of the oceans—a recent study suggests complete food chain collapse (Nagelkerken and Connell 2015)—water shortages in important agricultural zones and crop failures on a large scale (you cannot have genetically engineered high yield crops without water) and exacerbated by widespread monoculture, massive environmental damage and the loss of perhaps 1/3 of useable topsoils through erosion,[2] loss of organic and mineral matter and contamination by salt (Amundson et al. 2015), and subsequent loss of nutrition (for example Davis 2004) although this may be mitigated to some degree by the northward creep of agricultural production due to global warming (IPCC 2014), and global loss of biodiversity.

14.2.5 Space Development Timetable, Summary

Beyond the year 2040, there are no concrete plans drawn up for any space activity. As far as 2040, however, these projections above are rooted in what is currently being considered by players in the field. For example, NASA director Charles Bolden talked about the first Mars occupants arriving by 2040 during a question-and-answer session after his speech at Ohio University as part of an International Space University summer session, 1 July 2015.

It is instructive, however, to examine previous predictions about the speed of space development. The NASA study previously cited from 1977 thought that 10,000 people could be placed in a viable colony 20 years from the go-ahead. Jesco Von Puttkamer, responsible for long range planning for NASA in the 1970s

[2]It may seem like madness to speak of soils going extinct, but more than a third of the world's top layer is endangered, according to the UN, (which declared 2015 the International Year of Soils.) (UN 2014).

produced an evolutionary path tree diagram for these large space projects beginning from Skylab where a lunar base of 12 men was in place by 1995, a space station manned by 100 people at the same time, a Mars landing and permanent Lunar base manned by 200–300 people by the year 2000 (Brand 1977). Fifty years on, people living for long periods off the Earth number from between 3 and 6, occupants of the ISS. As a general operating parameter, numbers of occupants in off-Earth sites will be limited to the size of escape craft that can return them to Earth during a systems failure. It will be many years before properly established facilities with their own safety back-up systems can allow numbers of space centre occupants to grow without there being a means of escape for them.

Scientists, futurists as well as writers and cinema directors have all missed by many decades the points at which the particular phases of space development appear. One of the most important reasons for this is that the realism of the markets is never included or is ignored in these extrapolations. Many of the projected plans to exploit mineral resources in space make no sense to an earth-bound capitalist. Moon excavations and asteroid capture and mining are relying on the continued expansion of the Earth-based economy such that the rising Earth demand for metals such as titanium, aluminium, cobalt and manganese will pay for the development of off-Earth sourcing. The simple observation that whatever materials are discovered in space, the absolute minimum cost per kilo to Earth will be the cost of the launches required to retrieve this material which even the most optimistic of enthusiasts, Elon Musk, believe may come down to perhaps $2000 per kilo per launch (Messier 2014). Space activities are still not a self-sustaining economic sector and require continued government funding in various proportions to stimulate the private sector. There are no investment or hedge funds in rocketry, for example, even though rocketry is the essential component on which all investment in space relies. Returns on investment in satellite use are possible as long as the launch cost is brought down by government intervention, both currently and as a result of historic investment in space. There are no other space activities likely to provide sufficient returns for old-fashioned capital investments that are seeking higher and higher returns for less risk on its mobilisation, nor will the slow pace of space exploration provide material input into the Earth economy in sufficient time. The expectation that space exploration within the next hundred years will save humans by providing off-world reservoirs of individuals is way beyond what seems likely to happen in reality and completely ignores developments in robotics and AI which will be the preferred occupants of space for industrialists. Optimistic projections may have around 300 individuals working off the Earth in 70 years or so, and these would have to be able to sustain themselves without the presence of Earth to fulfil that expectation. The realities will be that most tasks in space will be performed by robots and most voyages will be automated. Human personnel will occupy bases or perform transitory maintenance in orbit but high numbers will not be needed.

Given that the occupation of space is humanity's long term objective, then clearly the preservation of Earth needs to happen first. It is therefore right to ask if capitalism is the economic means with which we can conquer space and build settlements in its own image since it has so compromised humanity's current situation. To answer this question let us first examine two components of the capitalist economy namely Capital and Labour and analyse whether the recent changes observed in the relationship between them will inhibit or stimulate space industrialisation and off world colonisation.

14.3 Part 3 Capital and Labour

14.3.1 Capital

> The question of the origin of capitalism may seem arcane, but it goes to the heart of assumptions deeply rooted in our culture, widespread and dangerous illusions about the so-called 'free' market and its benefits to humanity, its compatibility with democracy, social justice and ecological sustainability. Thinking about future alternatives to capitalism requires us to think about alternative conceptions of its past. (Wood 2002)

14.3.1.1 Feudal Fief, Responsibility and Money

While many thinkers have followed Marx in identifying the birth of capitalism with the feudal barons separating the peasants from contact and control of the agricultural resources (Marx 1867), this simple notion does not explain why capitalism is so wasteful and so disregarding of labour and the environment. The observation that owners of capital try to pass on costs of capital growth to others is not explained simply by the seizing of the ownership of the means of production by those who already had legal power over it. So let us briefly consider the feudal state anew.

A feudal settlement in the early years was a benign place, described succinctly by Mumford (1938), and confirmed in drawings of such medieval settlements all over Europe from the period. An outer wall enclosed houses and gardens, common orchards and grazing lands. There were common resources like wash houses and mills (in 1086, the Domesday Book noted 6000 water mills). There was clean water and sanitation. Very little waste; dung was returned to the soil. It was a barter society with very little coin. Studies such as have shown that the medieval population generally lived well with a higher than average calorie intake (Singman and McLean 1999).

After the collapse of the Roman Empire in the west, the amount of coin in circulation in Europe fell significantly and as a result, urban populations fell as people returned to the land (the Domesday Book Survey of England in 1086 mentions just 6 towns). In England, for example, the number of coins fell to a half between 900 and 1100 (Bolton 2012). The Feudal system of obligation began in a response to a lack of money. The hierarchy of duty and obligation passed from king to Baron and his knights to Lord of the Manor to serf and the produce of the land passed upwards in the opposite direction. While local centres of trade could flourish with barter, a year of poor harvest meant that some Barons could not supply the king with the expected produce; he was in debt to his obligation. Similarly if a

Baron was unable to supply the number of knights he was obligated to supply (through loss in battle or low birth numbers of suitable candidates) he was in debt. But since poor harvests were region wide, and knights could not be produced in any short time span, the only way the Baron could fulfil his obligations was through coin, either by giving it or using it to buy outside his fief. Thus coin began to be exchanged for obligation. Barons paid the kings in coin where they could not supply knights, and the kings bought mercenary soldiers with coin. Both the Church and the King demanded coin for projects not sustainable by local fiefs and so the feudal obligations were converted into money. With the monetary economy, coin represented a freedom from obligation to those who had it. Freedom from obligation meant that Barons became divorced from duties to the land and to their work force, a feature of capitalist economies, while debt replaced the previous social control of feudal duty, acquiring a moral dimension that it has to this day. Individuals found a freedom from obligation to their Lords in the urban setting, fuelling its growth, but which, being outside the barter system, also required coin. The importance of this was profound because as the population grew so did the numbers of free men who lived outside the barter system existing between serfs and their Lords of the Manor. Free men paid for their land use in coin, and Barons subdivided the amount of land under rent in order to get coin (for example, by 1298, long before the Black Death, 74 % of English tenant farms were under 5 acres (Dyer 2002) and so also reducing agricultural efficiency and thereby increasing the demand for coin even further. This interpretation is confirmed in the reforms of England's Henry II (1160–1170) which actually codified the separation of land title for freehold land from personal obligations, allowing a market for land to arise.

Coin was also needed to buy goods from far beyond the trading and mutual obligation confines of the Baron's fief so a demand for foreign luxuries also drove the need for coin. It is no coincidence that freedom, money and urban growth are interrelated and are indeed interchangeable. Where before the wealth of the Baron was measured in his labour force, labour freedoms became a drain on his coin, and his wealth was measured by what he could keep (memorably written down as demands in the Magna Carta of 1215). With the loss of obligations, Labour became a cost, and is accounted for as such in company balance sheets to this day. In this new function of money as a means of laying off responsibilities we find the birth of the moral power of interest, the classical separation between capital and labour and between freedom and the destruction of the environment.

If we jump forward to today, we can see this system of converting responsibility or duties towards the state and to the people in it into money as reaching its end game. In the economic crisis in the world economy nominally marked as beginning in 2008 with the collapse of Lehman brothers financial firm in New York, the strategy for fiat currencies run by central banks was to sustain asset growth but to keep money out of the hands of the consumer to avoid inflation. Early in 2015, there was $3.6 trillion of government debt around the world with *negative* interest rates (Moore 2015). Two-year government bonds are negative in Germany, Finland, Austria, Denmark, France, Holland, Belgium, Slovakia, Sweden, and Japan. It is worth noting that the Swiss government has become the first ever to issue a 10-year sovereign bond at a negative yield. While several European countries have sold government debt at negative yields up to five years of maturity —which means investors effectively pay for the privilege of buying it—no other country has previously stretched maturity this long. At the beginning of 2014, Bank of America said 56 % of global GDP is currently supported by zero interest rates, and so are 83 % of the free-floating equities on global bourses. Half of all government bonds in the world yield less that 1pc. Roughly 1.4 bn people are experiencing negative rates of return in one form or another on the safest places to keep one's capital.

Previously, rich economies had high levels of debt (though not as extreme) after the Second World War. In the UK, debt fell from 124pc of GDP then to 29pc by 1973, but has since been rising at a compound rate of 2 %—doubling every 36 years. There is now so much money from State issuance supporting assets in the form of deposit accounts that returns from assets beat any other kind of investment. This vast amount of monetary debt represents a freedom for the class of asset holders that has perhaps never been seen before. Following our reasoning, *negative* rates of return should indicate a reversal of the moral power of the issuing authorities whereby those organisations and non-state actors that accept such negative returns from the State acquire greater moral authority over it: they are, in effect, buying more control and demanding greater power and freedom from sanction in economic negotiations. This state of affairs is represented in the secret economic agreements currently being negotiated in which enterprises can force changes in government policy and local laws such as labour union representation where such policy and law reduces profits (for example the Transatlantic Trade and Investment Partnership which is considered by some to be anti-democratic for this reason). As States lose their authority over national economic interests, they lose the power or right to divert resources to the capital intensive low return space activities we considered in the space development timetable, lengthening achievement dates still further, and generally braking long term investment in the 'common weal' of space industrialisation.

This creation of debt to support assets reflects a systemic change in the utilisation of capital. In the past, banks lent out from their cash deposits against the value of an asset or promise of a later but higher cash return. But with government purchases called 'Quantitive Easing', an asset is created out of a liability by the central bank who takes the liability of a favoured bank (usually arising from buying government debt but not always) prints money and creates a deposit in a favoured bank in return for the debt. The favoured bank then uses this deposit as the basis for loans that increase asset values or purchases of more debt, but not to go into production for that would raise the inflation rate. As a result banks loan to companies who buy back their own shares; the earnings per share automatically rises and by shedding workforce, the productivity metrics also rise, even as there has been no actual contribution to the overall economy. Assets purchased with the central bank cash, at low risk because interest rates are low and the Central Bank is the buyer of last resort, rise in value in the market. The existence of this debt, however, is a real danger to the health of the Earth's biosphere since for repayment to be possible, it requires eventual growth in consumption, in inflation and in waste to reduce the burden of cost on capital. The cheap interest rates are thus divorced from the real risks of investing (where returns are supposed to reflect risk), while the loss of consumer power also restricts investment in job-making industries. (This process can be confirmed in other ways, for example, where we see company bond yields rise well above those of government, and where pension and insurance funds fail to reach historic rates of growth.) Such returns are not present in the foreseeable space industrialisation timetable and as a result, investment in space activities will fall further if this current economic situation continues.

The risks in new industrial investment cannot compete with risk-less asset value gains resulting from the strategy of the central banks. As asset values rise, the only business that makes any long term sense is to live off rent or to invest in the new-tier economies that don't employ much labour. And thus we are drawn inevitably into differentiation and inequality in the consumer economy and this has an impact on the capital available for space investment.

14.3.1.2 Inequality

The wealth inequality in the world also nicely confirms the turning away of capital from industry. The concentration of wealth into the hands of a few seems to have produced none of what orthodox economists expect from surplus capital which is to be an engine of growth and innovation. The inequality is not the result of innovations or entrepreneurship of the highest order—although of course there are some true entrepreneurial individuals in the top say 400 (who have doubled their share of wealth in the last five years). Nor is it a pure consequence of stratospheric salaries paid to executives. This inequality is derived from *rent*.

> ...In 2014 there were 1,645 people listed by Forbes as being billionaires. This group of people is far from being globally representative. Almost 30 % of them (492 people) are citizens of the USA. Over one-third of billionaires started from a position of wealth, with 34 % of them having inherited some or all of their riches. This group is predominately male and greying; with 85 % of these people aged over 50 years and 90 % of them male. There are a few important economic sectors that have contributed to the accumulation of wealth of these billionaires. In March 2014, 20 % of them (321) were listed as having interests or activities in, or relating to, the financial and insurance sectors, the most commonly cited source of wealth for billionaires on this list...Between 2013 and 2014 billionaires listed as having interests and activities in the pharmaceutical and healthcare sectors saw the biggest increase in their collective wealth. Twenty-nine individuals joined the ranks of the billionaires between March 2013 and March 2014 (five dropped off the list), increasing the total number from 66 billionaires to 90, in 2014 making up 5 % of the total billionaires on the list. The collective wealth of billionaires with interests in this sector increased from \$170bn to \$250bn, a 47 % increase and the largest percentage increase in wealth of the different sectors on the Forbes list. (The Oxfam issue briefing 2015)

This briefing lists a mere 9 % of the list having acquired their wealth from industrial companies. The Harun web site that keeps watch on wealthy individuals attributes the wealth of a mere 11 % of billionaires to technology (Hurun Research

Institute 2015) The rest lack direct responsibility for their wealth accumulations. The richest sector of society is living off the interest from a narrow range of other people's activities and not from industrial production or their own entrepreneurship in technology. In addition the global expansion of the middle class (Kharas and Gertz 2010) will not only intensify world consumption, but will strengthen a two-tier market with asset-less 'serfs' at the bottom and a higher income band of consumers supporting the growing assets of the elite (Schor 1999).

14.3.1.3 Changes in the Movements of Capital

Changes in capital functions fall into two parts: capital mobilisation and the distribution of holders of capital.

Capital Mobilisation

We will consider a single metric, the ratio of market capitalisation to employee in enterprises. As the simple plot below shows, old fashioned enterprises where workers interact with the real substantial world remain limited in their ability to increase the returns on capital employed in the enterprise. There is a new economy based on information that is learning how to make use of both the consumer's capital endowment as well as the dark capital that individuals possess. It is yet another way of transferring enterprise costs away from the capitalist and to the common tax pool by cutting down on workers and what they have to be supplied with to perform their functions. I call these enterprises the new-tier economies and they support increasingly fewer households globally which can be seen in table 14.2.

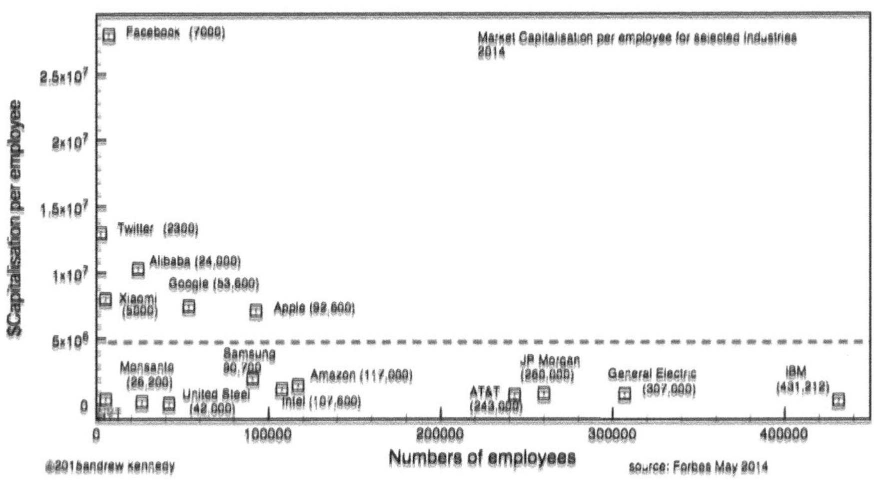

Fig. 14.2 Linear plot of market capitalisation per employee against numbers of employees for selected companies. Data source: Forbes May 2014

Traditional enterprises all fall in the MCPE band of between 1×10^5 and 5×10^6 dollars of market capital per employee. Each enterprises employs thousands of people and manipulate real objects. They are engaged in material society in a much greater way that the new-tier enterprises. They are, as it were, providing for many households in the economy. Wallmart (off this chart), by this metric, does an enormous amount of social good since it mobilises capital to provide for 2.2 million households, giving it a MCPE at the bottom of the band of 1.12×10^5.

But when we look at the new-tier enterprises like Twitter and Facebook we see the opposite trend. Enormous amounts of capital mobilised to provide for very few workers.

On this plot, Uber is over to the left and way above even Facebook. Its market capitalisation of $40 bn sustains just 1000 employees, with an MCPE of 4×10^7, similar to the crowd labour enterprise of Instacart. Instagram had an MCPE of 8.3×10^7 when bought by Facebook for $1 bn. But this pales in comparison to Whatsapp which, with an MCPE of 34×10^7, supported just 55 employees when Facebook bought it for $19 bn.

It is instructive to apply this same metric to NASA. If we consider NASA as an enterprise with its 18,000 employees and 40,000 contractors as its labour force and consider its yearly budget of $18.3 bn (in 2015) as equivalent to its earnings, and if the 1:8 stimulus multiplier commonly assumed can be applied to its earnings to derive its total value in the market place we find a typical MCPE of 2.5×10^6, in the middle of the band of traditional enterprises. Equivalent to AT&T or General Electric, though supporting fewer households.

And if we look at the MCPE ratio of Space X, at 3.0×10^6, we can see current space industrialisation cannot compare with the new-tier economy. Mixed industries involved in space activities have an MCPE of even less, such as Boeing, at 6.5×10^5; Lockheed, at 5.6×10^5. Indeed these figures are so low that one is tempted to speculate about how much longer these two companies can stay involved in the space economy without continuous government support (other enterprises dedicated to space activities are private companies and figures are not available). The conclusion is that NASA's activities in the long term are not especially beneficial or attractive to modern investors, and, by direct comparison with space industry enterprises, that there is little reason for capital currently to be drawn into making rockets or investing in planetary missions.

These new-tier enterprises are making use of a new phenomenon to get their high MCPE ratios, the mobilisation of the 'dark capital' of consumers. It is in the utilisation of dark capital that we see the consumer economy separating into two levels of engagement, with implications for space investment. These new tier economies are leveraged on a specific level of asset accumulation in our consumption, and on what is euphemistically called 'big data', whereby individual consumption is mined for specific behavioural characteristics which can then be applied as sales techniques (see also, Odlyzko 2003). For example, Facebook's business model relies on its users' purchase of a computer or mobile device, but it also relies on a user's phone contract, and contract with a server network; it needs not only a user's personal data, but that of friends and family and the user's efforts to connect with people and finding and posting material, including raw witness images of events, all of which Facebook profits from. It needs a user's time, judgements and tastes, a whole life. Facebook doesn't have to have journalists in the field, information researchers, graphics departments cropping photos, sub-editors or any of the workers like one might find say in a newspaper office. Its effective workforce consists of programmers and a support department, the material input is all supplied by the consumer.

An individual also has yet more dark capital in his or her state of health, in his or her genome, his or her immune system, his or her gut microbes, his or her organs and a nervous and perceptual system that can be connected directly to machines or computing devices, none of which he or she owns the instant any cells, organs or limbs items have left, or been removed from his or her body (not including their whole DNA which is natural resource and cannot be patented). The era of the transhuman heralds another shift in the worker/capital relationship where labour and capital become intertwined and the individual no longer exists as a separate independent worker 'soul' with its own rights but is owned in certain degrees by whomever initiated the transhuman changes.

In this new economy of the dark capital exploiters, where consumer behaviour is the essential data, the exploitation of dark capital by enterprises must necessarily move up the consumer earnings spectrum for their investment to make long term sense. There will be privileged consumers and entry point consumers and the experiences of each will be different because the earnings power of each will be different. Entry point consumers will belong at the lowest end of supply and will only provide a limited amount of interest to the new-tier enterprises, which means that not only will this group have access to poorer goods but also to poorer services, resources, education, health and the rest. Facebook has already put forward the idea of a two-tier internet so that internet usage can be more closely tied to consumer behaviour (the Free Basics program). Internet server companies have been trying to instigate a two-tier internet access for some years,[3] and many server companies informally 'choke' off demand from non-enterprise consumers when traffic increases, creating a de facto privileged user. This then, highlights the problem for those companies wishing to provide internet services through satellites. Half the world remains to be connected to the internet and yet this half, mostly rural populations, do not yet purchase enough goods or services to develop their dark capital and provide the promise of further consumer involvement.

Distribution of Capital Among Owners of Production

But even among capitalists there is a concentration of power in fewer hands and less sharing of the returns of capital among the asset owning classes. We have seen how

[3]The European Parliament now supports payments for specialised internet services as long as open access is not compromised according to new net neutrality rules (European Commission 2015).

the new-tier enterprises support fewer households than traditional capitalist enterprises by exploiting human dark capital. But even among capital owners there is very much less sharing of capital returns with investors through shares and other corporate democratic means. A recent Economist report explains that,

> …In business, too, family companies continue to thrive, as our special report in this issue explains. More than 90 % of the world's businesses are family-managed or controlled, including some of the biggest, such as News Corp and Volkswagen, a carmaker in the throes of a boardroom battle between its two main family owners. The Boston Consulting Group calculates that families own or control 33 % of American companies and 40 % of French and German ones with revenues of more than $1 billion a year. In the emerging world the preponderance of family control is greater still… (Economist 2015)

The Economist study study found that the richest ten families controlled 34 % of market capitalisation in Portugal and 29 % in both France and Switzerland. In a previous study The Economist noted that around 85 % of $1 billion-plus businesses in South-East Asia are family-run, around 75 % in Latin America, 67 % in India and around 65 % in the Middle East (Economist 2014).

Some of the worlds great multinationals are family controlled like Walmart, Mars, Samsung, BMW, Ford. Foxconn and many more. VW is 51 % owned by the Porsche family and its chairman Mr. Piech. Even Warren Buffet is grooming his son and grandson to continue the Buffet Enterprise culture. Private companies also enjoy similar benefits. Dell, the paragon of schoolboy entrepreneurship took back public ownership of the company he founded so he could better manage it as a private concern.

Family run enterprises by-pass the problems of the share-holding democratic management where secrecy and compromises are open to public scrutiny. Publicly traded companies more easily mutate and change from their original conception. This is especially true of companies started by inventors. Whereas family run enterprises can maintain their desired focus regardless of cost. Decisions about how much to take out of the company can be kept secret and they can avoid take-overs and attempted coups by shareholders (internal feuds notwithstanding). They can play a long game and not worry about shareholder needs. (The world's longest-lived enterprise is a family run bank, Monte dei Paschi di Siena, headquartered in Sienna, Italy, which has been operating continuously since 1472.) They answer to nobody but themselves and often invest more in reputation than in plant and innovation. In this way at least, family-run enterprises mimic the feudal barons and fief holders of old. Just as capital is no longer being spread around labour, a share in the ownership of capital is no longer being offered around potential capitalists as much as one might have expected. The lack of broad representation of capital owners in the management of enterprises is also figured in the basic inequalities between sectors of the economy, and family dynamics are also beginning to play a significant role in governments. Even in the financial corporate world, however, a shadow banking system has arisen in which many unregulated banking functions are done outside the markets and across national boundaries not by banks but by hedge funds, pension funds, securities brokers and other kinds of investment vehicles that usually take assets rather than monetary deposits as the basis of deals. The value of the shadow banking system is thought to be as much as half the traditional overt system, and plays a hidden role in determining asset values. It is specifically designed to create leverage without regulatory oversight. Again the system reduces the orthodox sharing in capital, and the utility of true risk.

It is well-known that family dynasties play a role in the politics of underdeveloped countries, undermining all attempts to open up the political process to democracy and to reduce corruption. For example, in the Philipines 70 % of Congress belongs to dynastic families (Yousingco 2015). In India, it is said, every member of Congress under 30 years of age belongs to a political dynasty. The dynastic idea is so strong in India that when Rajiv Gandhi, son of Prime Minister Indira Gandhi, herself daughter of Jawaharlal Nehru, the first Prime Minister of India, was killed, his retiring wife, Italian-born Sonia was forced to become head of his Indian National Congress party, and her son is now heir apparent. The US has seen the Roosevelt and Kennedy dynasties, and the next US presidential election in 2016 may well see a battle of the Clinton/Bush political dynasties. Hilary Clinton is the wife of former President Bill Clinton, and a Republican hopeful, Jeb Bush, is brother to one President and the son of another. The Clinton's daughter, Chelsea, is the vice chair of the influential Clinton Foundation and is working in Hilary's Presidential campaign.

A brief list of recent leaders around the world includes: in the United States, a republican former president George W Bush is son of former president George Bush; in Argentina, president Cristina Fernández de Kirchner is wife of former president Nestor Kirchner, (and let us not forget the earlier Peron attempts there); in Japan, former prime minister of Japan Yukia Hatoyama is grandson of former prime minister Ichirō Hatoyama; in Thailand, prime minister Yingluck Shinawatra (recently deposed) is sister of former prime minister Thaksin Shinawatra; and in the Philippines, former president Gloria Macapagal-Arroyo is daughter of former president Diosdado Macapagal, while the current president, Benigno "Noynoy" Aquino III, is the son of the former president Corazon Aquino; In Malasia, Prime Minister Najib Tun Razak is the son of the second prime minister, Tun Abdul Razak: in France, the current president, Francois Hollande is former partner to Sergolene Royal head of the socialist party and former Presidential candidate. In Canada recently elected Prime Minister Justin Trudeau is son of former Prime Minister Pierre Trudeau.

Family continuity in politics and economics is a strong indicator that democratic movements are faltering, where economic and political power are no longer being shared in true democratic fashion, and where, in practice, we are entering a realm in which fiefs dominate the economic decision-making process of the State. An additional indicator of this loss of State authority is the idea previously discussed of corporations gaining a moral ascendancy over the State as a result of low or zero interest rates. A third indicator is also given by the growing involvement of the richest individuals in political organisations. Not only do the rich fund political candidates and their organisations, they head lobbying organisations to steer decision-making their way. During the run up to the 2016 US Presidential elections NY Times researchers revealed that just 158 families had provided nearly half the early money of the candidates campaigns (Confessore et al. 2015).

14.3.2 Labour

Since the end of Feudalism and the beginnings of the industrial revolutions, where resources came to be wholly owned and workers could be divorced from the bounty of the land, a worker in an enterprise has been accounted as a cost. Workers are a cost such that even where the balance of income to cost of an enterprise is zero (that is where all wages are paid for), the enterprise is considered not profitable for capital. An enterprise's natural disposition is to increase profitability, not to create jobs. The unemployment dilemma is revealed in government action that tries to create jobs through spending and other stimulus and to simultaneously make capital more profitable (the private sector), and is undoubtedly the most confusing problem for economists. The numbers of workers, with their rising health care and other non-work related costs; who represent a huge percentage of business costs and overheads, need to be cut for companies to increase profits. Productivity and profitability all imply cutting jobs where possible, and certainly in static or slow growth situations. The nascent two levels of consumer can be found emerging in the period of time, 2008–2015, where low interest rates and reduced consumer demand allowed companies to buy back shares and lay off labour at the same time, producing improved earnings and productivity without investment. Higher share values enable take overs and thus reduce competition between enterprises further deflating wages. Price inflation is rising while at the same time the purchasing power of the lowest levels of labour has shrunk and its wage negotiating power has almost disappeared.

This situation for labour has been described quite forcibly by Standing (Standing 2014) where the global labour market will suffer from insecurity, downward pressure on wages and the loss of entitlements and protection, and this is in spite of the recent fall in populations, (and thus in workers) generally lauded by economists as a relief from Malthusian disaster. There are a number of deflationary pressures on the consumer economy which increase the imperative of capital to find activities that will provide a return from the consumer sector that retains purchasing power, and thus making a long term investment in space activities less attractive.

14.3.2.1 Urban Population Growth

All over the world national populations are growing less fast, although as shall be seen, the reasons for this are not clear. In many countries the overall fertility rate is below replacement. One of the current explanations for this is urbanisation, where the labour of children cannot increase the productivity of parents as much as they can cover the children's extra consumption in an agricultural setting. There are family businesses of course that can do that, but in general a typical city worker cannot afford a big family. This ignores the fact that in many rural societies, agriculture is already at the margin and cannot increase productivity thereby forcing people off the land rather than absorbing them. For example, In India, the average size of small farmer holdings is being continually reduced. Of the total land holdings in 2000 nearly 63 % were operated by farmers having less than 1 ha of land (2.3 acres). It is also estimated that farmers with less than 4 ha are not financially viable Singh (2009) such that agriculture does not provide for population increase as expected and thus urbanisation becomes more supportive of population growth, not less. As A. Darrat, and Y. Al-Yousif, unravelled in their paper (Darrat and Al-Yousif 1999), population growth can be positive or negative depending on economic freedoms available especially on the rights to private property and to engage in trade with minimal interference. So the cause of local population growth rate decreases tend to be due to problems with local land ownership, property rights and productivity rather than the long term overriding trend in fertility.

This has happened before. Similar processes ended the Feudal period where gradual urbanisation was the symptom and encourager of population growth. As the Roman era ended, people left the cities and went back to the land because the cities lacked money. Life could only be sustained through agriculture and barter. During the period from 1000AD to 1400AD, urban centres were actively created all over Europe as centres of trade and points of distribution, while more and more land was brought under agriculture. The city state flourished particularly in Italy and Germany and was an important engine for the subsequent Renaissance. During the later feudal period where feudal agricultural productivity remained low the continual subdivision of agricultural plots due to higher populations failed to sustain families living on them. Free farmers were absorbed in urban centres which also gave an opportunity for bonded labourers and serfs to find their freedom and to engage in economic activity, and provided places of refuge from the continual turmoil of the later Middle Ages.

Even 400 years later as the industrialisation of the cotton trade produced large factories in England, population growth allowed urban centres like Manchester to draw people in because the mills employed women and children and whole families could find work (Beckert 2014). Manchester alone, at the height of the cotton boom employed nearly half a million workers in its factories (almost 10 % of the country) and was the inspiration for Malthus's work in the early part of the 19th century. Between 1700 and 1800, England's population grew from 5 million to nearly 9 million, so the excuse that urbanisation now is the principal cause of general population *decline* is a partial one at best. China's urban growth, for example, has been put at 2.5 % p.a, as it makes the transition from a predominantly agricultural society to a predominantly urban one. By 2050 it is expected that 2/3 of all humans will live in cities and this will impact on population estimates.

Where there is urban decline, mostly in the developed world, we can see this as an example of Zipf's Law, a general power law (plotting rank vs frequency), which states that in freely competing populations, the most populous city will be twice as big as the second, and that twice as big as the third and so on. This holds for those cities that are better integrated into their environment. Smaller cities with less integration will tend to leak population to the larger better linked ones. Which agrees with historical changes at the end of feudalism that we have mentioned. The conclusions are, supported by recent studies, that far from peaking at around

9 billion in 2050, the world population will continue to grow to 11 + billion by the end of the century (UNDEA 2015).

Urbanisation growth has another consequence in predominantly rural societies in opposition to the reason that children are less required as they cannot increase family productivity as easily as they can in the rural setting. As agricultural workers leave the land, they put downward pressure on urban wages of working parents, the pressures on children to join the labour force to improve urban family fortunes is increasing. As wages stagnate, child labour rises and forms of slavery and bonded labour are increasingly putting children to work around the world both in cities and in agriculture.

14.3.2.2 The Growth in Slavery

Slavery had a good run in the past and it seems to making a comeback. Both ancient Greek and Roman cultures were built on it. It wasn't until the 1770s that the movement to abolish slavery took hold. The British Parliament abandoned slavery in the British Empire in 1833. Serfdom was not eliminated in Russia until 1861, around the same time a civil war was fought in the US to eliminate slavery there (the constitutional amendment to make it so was passed 1864–65). Brazil ended it in 1871, but informal slavery carried on in parts of Africa while modern political enactments of feudalism like communism and its variants sprung up around the world, where individual rights were completely subordinated to the political will of the state.

Yet, the ILO estimates that modern-day slavery is $150 billion per year business, and 20.9 million or more people are working as modern-day slaves, victims of forced labor, trapped in jobs into which they were coerced or deceived and which they cannot leave. This figure is a conservative estimate, given the difficulty of measuring this disguised crime that includes an estimate of human trafficking and sexual exploitation (ILO 2014).

The question of whether slavery was profitable is long discussed. When we consider the question of dark capital we can see that the entire trading value of the person is his or her dark capital. A person buying a slave is buying not simply labour but the skills and knowledge such a person may have, the ability to pass them on as well as to reproduce. In addition, the dark capital includes the will and instincts an individual may have to survive in adverse conditions. Children as slaves have much less dark capital and as a consequence their price is very low. In 1850 it has been calculated that a good adult slave could fetch as much as $40,000 in today's money. Whereas today a child slave can be found for a little as $90 (Willamson and Cain 2011). Slavery is important to include in our argument because it depresses the value of labour and reduces its capacity to extract value from capital but which leaves capital even less able to gain returns in a consumer economy.

14.3.2.3 Demographic Time Bomb

It's not only absolute numbers of consumers that matter but also the proportions of ages. Deflationary pressures come from the fact that the working population is ageing and while half the world's population may be under 30 years of age and over a quarter are 15 years old or younger, the numbers of workers has peaked. Some developed countries like Japan (where 30 % of the Japanese are over 65, topped only by Finland) have too many old who demand high standards of living, while others have too many young who need to be fed but are not yet productive.

As an example, in a mere 20 years the ratio of worker to retiree in the UK has gone from 3.15 to 2.6 and it is expected to continue to fall. Quite how a youthful population can support both a quality retirement of the aged and the consumption of the new generations when wages generally rise to a peak late in life and as the availability of resources is also declining, is, at the moment unknown. It may require a different economic system, although it is hard to envision one that maintains the benefits to holding capital as well as encouraging technological development.

From now on there will be increasingly more people to be looked after than there are people to work to pay for them. Fewer workers should mean higher wages but more retirees requires a higher return on capital, which tends to depress wages, lowering consumer power. The massive debt burden of governments can only technically be paid off with rising inflation but this is difficult to manage as the numbers of workers and their total consumer power decreases.

14.3.2.4 The Shrinking Work Force

The labor force is shrinking in many countries. The US labor force participation rate has declined since 2007 as has the civilian employment to population ratio (USBLS 2014). The numbers of individuals participating in work in the US has fallen from a peak of over 67 % of the population in 1998 to 62.8 % at the end of 2014 and is not projected to improve. The 5.7 % reported US unemployment rate is achieved by not including discouraged workers as part of the work force. (A discouraged worker is a person who is unable to find a job and has given up looking.) The US government stopped including long-term discouraged workers (discouraged for more than one year) in 1994. If the long-term discouraged are counted, the current unemployment rate in the US stands at 23.2 %. A second official unemployment rate, which counts short-term (less than one year) discouraged workers and is seldom reported, stands at 11.2 %. Not surprisingly, wages have not risen in the last decade.

Not only is the world's labour force shrinking both through demographics and through discouraged workers who exit the workforce and don't come back, the new-tier economies don't need as many workers as they once did. Technology, through the computer, has given capital a chance to reduce the claims that labour traditionally make on capital both by creating enterprises that exploit the individual dark capital of humans and by those that eradicate workers with automation and sourcing employment by the hour.

Furthermore trends of land purchasing in many parts of the developing world are turning agricultural workers into migrant workers with only seasonal work and no stability or opportunity for advancement as agriculture looks to increase its return on its capital. These classic 'land grabs' as they are called are disenfranchising land users and, in Africa and South America, sometimes privatising areas of communal use (that have the undesirable effect of keeping poverty-driven population up but economic growth down). Changes introduced in India this year will permit precisely the same concentrations of agricultural land in the hands of enterprises that will mechanise production and reduce workers. Some areas of Africa land is leased to businesses by governments who have also newly created rights to a water supply, thus further displacing farmers from their local and ancient arrangements. These 'clearances' have the same ring that the clearances and fencing off of common land had in the 16th century in England, although that process had already begun earlier in the 12th century. While higher production may be the target of such deals, according to Oxfam, at least 60 % of produce grown by international concerns is exported and fears of monoculture ruining ecosystems for the benefit of the export trade are a genuine concern.

14.3.2.5 Automation

Continuously advancing technology is disrupting many of the very industries it helped build—and in many cases, those rapid advances are proving to be real net job killers. Traditional enterprises that cannot derive increasing returns from their capital are driven to automate the workforce. Martin Ford has described the disruptions sweeping through industries like publishing, where robots can write and deliver news stories on-line to web sites, whose sole job is to capture readers' interests and to display advertising, and also in music, retail, and manufacturing, health, and even in higher education, where studies show that robots can score essays in exams better than humans (Ford 2015). Robot tutors are taking over on-line education and will undoubtedly supply almost all undergraduate education in the near future. Robots are short order cooks putting at risk millions of advantageous jobs in fast-food outlets. They are diagnosing health problems like cancer or dementia, mixing the pharmaceuticals for them and administering them. They will drive delivery trucks and taxis and almost certainly will be the only permitted vehicles on dense urban motorways before long.

3-D printing is expected to reduce employment in many normally labour intensive industries like construction, while at the same time giving a few new opportunities in areas like design and personalisation of output—sometimes called the market of one. There is no way to predict the social consequences of this technology except that loss of jobs in the near future in many sectors of the economy is assured. In earlier decades, just-in-time information has been reducing the need for warehousing staff and the wastages of stocks of parts, and now the 3-D printing revolution is reducing them even further. Very few manufacturing systems would be immune from the automation revolution, as still another trend appears on the horizon, distinct from automation, that of biological manufacturing systems, creating not only drugs and foodstuffs but also human body parts, protheses and plastic objects.

The interconnection of devices—the Internet of Things—will also become more widely used in industry and manufacturing in order to streamline manufacturing from design to delivery. Developed economies will benefit the most from new technologies, which will destroy some jobs and create new ones, but developing countries will suffer as the need for cheap labor in low-end manufacturing declines. Cybersecurity will become a greater problem as "smart" technologies become more integrated with manufacturing. People will be needed less by armed forces and even the ordinary soldier that is employed will have to have higher skills in information technology reducing the need for the basic soldier—the grunt—of armies, and the loss of a route out of poverty for many.

Along with automation, a secondary effect of the information revolution is to enable individuals to exploit their own accumulated resources for gain. We have discussed a consumer's 'dark capital' previously but we can describe another trend in the growth of mobile and information technology allowing for the employment of low grade labour, namely, zero hour contracts, just-in-time distribution which does away with warehousing and even packaging, and crowd-sourced labour, where groups of dispersed individuals are organised to perform a service. Science projects have been using crowd labour to work on distributed processing for a while now, like searching for planets or comets or nova in stars, or using individuals' computer's rest time to compute functions. Now we also have commercial enterprises connecting with these group capacities. For example, Instacart, a shopping service employing hundreds of people to do its users' shopping, ordered online, and to deliver the produce using their own transport. It was recently valued at about \$2 bn with only 50 full time employees, not counting the crowd labour. This is a market capitalisation per employee of 4×10^7, about equal to Uber. Traditional work forces are now seen to require so much management in-house, taking too much time, space and organisational effort that the minimum wage work force has mutated into a minimum hours work force, further depressing the capacity of labour to participate in the consumer economy to the fullest.

14.3.2.6 Loss of Representation

Unemployment rates in America, Britain and Japan—all of them at or below pre-crisis lows—rose to high levels between 2008 and 2014, and would previously have triggered rising wages. The fact that it did not is a testament to the success of years of government legislation undermining the power of unions. Collective bargaining has almost vanished from the developed countries, which is curious since there is a wage premium for union membership—around 7 % for UK workers (UKONS 2015) for example. 20 years ago 60 % of wage levels arose out of collective bargaining in the US, now only 6 % of wages paid are the result of collective bargaining. In the UK union membership has halved since 1979 and trades unions membership in the private sector is now 14.2 % of the workforce though a higher density than the European average (UKONS 2015). In the US in 2010, union membership was 11.4 % of workers and a mere 7 % in the private sector, while in 2014 total full and part time workers representation by unions was 6.6 % (USBOS 2014). Some countries e.g. Saudi Arabia, still have no unions at all. Where there is growth, however, it has been in insecure forms of employment: part-time work has risen, as have the ranks of the "underemployed", who would like more hours if they could get them. As looser contracts have helped create flexible workforces, casual work—from drivers for Uber to day labourers in construction—has boomed. Temporary jobs may be up, but workers' bargaining power is not, further depressing wages. This pattern is repeated around the world. Ironically, the lack of union power has led the shift into self-employed or autonomous workers which means a problem for governments who need regular and consistent employee contributions to fund pension programs. Governments will undoubtedly start to control the levels of cash in the economy further disadvantaging the lowest levels of labour who rely on the black economy to survive.

14.3.2.7 Part 3 Conclusions

At the start of the Feudal period and through the industrial revolution labour was the means of realising wealth for owners of capital. This is no longer the case and workers tend to be rewarded only in so far as they contribute 'capital' of their own. Professor Angell of the LSE is one who similarly predicts that society will come to be strictly divided between those contributors to society who possess assets (like sufficient dark capital) and from whom wealth can be created and those without assets and who will receive no or few social benefits (Angell 2000). Capital has found a way to use information technology to increase returns from capital while depressing labour involvement at the same time. These changes in labour support the recent changes in the way capital is employed in the economy and appear to be systemic rather than temporary. Earth's population is slowly being divided into those who can participate in high level consumerism and a labour force that remains outside this economy and is divorced, in the main, from opportunities to enter it. There are two forces at work concentrating the differences in each sector, the increasing failure to employ capital in ways that distributes the benefits of its mobilisation among the population, and demographic changes and loss of negotiating power in the labour force. Labour has less and less power to demand a share in the returns from the mobilisation of capital and it is unlikely that it will be able to acquire this in the future without radical political changes. In conjunction with the loss of agricultural tenancies, loss of menial and factory jobs, the loss of cash economies and the rise of automation in all spheres, a further trend is the trend towards the transhuman worker who will give new meaning to labour, especially in the space environment, and to the traditional marxist class divide by being both an expense to an enterprise and also a capital asset. This additional separation of labour from capital is well documented in the wealth inequalities of the present day.[4]

The well-off young gain even more benefit from the inequality in education and opportunity. The gap in test scores between rich and poor children is 30–40 % wider than it was 25 years ago, firmly fixing the inequalities in successive generations since the intangible social value derived from privilege in education makes social mobility even less accessible. Migratory and immigrant workers tend to lack education for the surviving blue and white collar jobs and this coupled with social prejudice suggests that large pools of unskilled workers will gather at the bottom levels of the social pyramid with little opportunity to rise into the middle classes, and as a consequence will be of much less interest to capital and to the new opportunities for capital growth.

We can see that changes in labour populations, changes in the way capital is employed with less labour than previously, and the loss of broad ownership of capital in conjunction with loss of authority of the State over the deployment of capital is already leading to a narrowing of goals and shortening of horizons to future economic development here on Earth and lengthening still further the space development timetable while climate change and biosphere damage will continue to undermine the space economy and the risks capital needs to take to invest in space. By the middle of the century, space activity will not have advanced while Earth economy may still be mired in difficulties. The idea of 'escaping' into space seems morally reprehensible under these conditions.

14.4 Part 4 Space Colonies

14.4.1 The Space Economy Dilemma

Having gone into some detail about critical factors to the delay in the space industrialisation process up to at least 2050, we can see more clearly that the capability to colonise and to establish off-Earth settlements is far in the future. Given the rate of development we have outlined, where colonisation is unlikely to occur unless the damage to the Earth's biosphere is minimised and solutions for a healthy, growing but sustainable consumer economy are in place, it is highly likely that the methods of colonising space will be part of the solutions. But where will these solutions come from? NASA still believes that its competitive funding model will stimulate the private sector to take up more and more of the research and manufacturing load yet the competitive model and the competitive impulse between

[4]"GDP growth isn't doing much to raise your income anymore. And the trend seems to be getting worse: since the 2009 recovery started, 95 % of GDP growth has been captured by the top 1 %. Under such conditions, if you are not among the top earners in America, you may not care very much whether the BEA announces that the economy grew at a 3.6 or 2.8 % pace in the third quarter." (Economist 2013).

nations for space resources and territory belong with the extreme capitalism that is currently in crisis and looks increasingly unable to sustain the pace of space development. Further, the surplus of debt created to produce industrial production in space will undermine the very market it needs over time since it must be Earth that supplies the markets for space industrialised production as well as the capital, of which a great deal will be lost to the Earth economy by way of structures, spent rockets and fuel and other material being left in space.

Corporations and nations seem to have already recognised that the space economy is not working and, as we might have predicted from the earlier analysis, in requiring less risk and shorter timescales to their returns, are giving all the classical signs of subverting political processes by undermining past UN treaties and agreements that tend to contradict profitability in a space venture. For example, attempts are being made to undermine a 20-year old UN agreement to share internationally meteorological data by US enterprises wanting to receive a return from the launching of a new generation of weather satellites. The UN Committee of Peaceful Uses for Outer Space set up in 1959 recognises that a number of issues need urgent attention such as space debris, nuclear power in space, the filling up of the geostationary orbit and other issues, yet no agreements on these issues between nations have been reached. In spite of the US being a signatory to the Outer Space Treaty where nations agree that the development of space should be conducted to the benefit of all nations whether or not they are space-faring nations (matters of defence are not included), the US Federal Aviation Authority has given the go-ahead for US enterprises (like Bigelow) to own installations on the Moon, and US legislators (under intense lobbying by Space Resources and Deep Space Resources) are introducing the Space Resource Exploration and Utilization Act of 2015 to give anyone who captures an asteroid ownership of it, and indeed to give a probe a zone of 'ownership' 125 km radius around it as this is the zone of the right to non-interference referred to in the Outer Space Treaty. If this US act remains unchallenged in the International Court then it effectively turns any manned solar system landfall into a territory grab for any nation, and we can look forward to Russian, Chinese, US nation territories on the Moon by 2040 and with their separate legal systems in operation. The US national space policy is committed to the principles of a "robust and competitive industrial base" in developing space (The White House 2012). One expects that every nation will have a similar attitude.

Supporters of this essentially capitalist manner to space development underpin it with the classical model of exploration and risk and which many commentators think will characterise space settlements. Two significant features of life on Earth are generally ignored in this model, however. First, space is not analogous to the surface of the Earth, and second, growth on Earth relies on free biosphere productivity. Taking these facts into account gives us a different background to space colonisation.

14.4.2 The Biosphere Productivity and Human Exploration

An obligation to obey the human exploratory drive is often quoted as one of the reasons space should be colonised. It is certainly a fact that the human species moved out of Africa and spread out over the surface of the Earth to occupy almost every habitable region and even inhospitable climates like the arctic, and deserts. Such expansion was done in a complicated way with groups overlapping, traversing the ground at an average rate of around 1 km a year, hardly an explosive rate of exploration by humans who are claimed to have a genetic disposition to move at whatever cost. Space exploration is not analogous in any way to this expansion. Thanks to the Earth's biosphere almost all voyages, whether on foot, by sea or land, were made over a productive surface that could sustain the explorers with little cost. Deserts may be the most inhospitable regions on the Earth, but Bushmen were quite able to survive in the Kalahari deserts of their ecological niche. The open oceans are not equivalent to space, and are, or certainly were, teeming with life. Captain Bligh, abandoned in an overcrowded open boat with few provisions, sailed almost 4000 miles across the Pacific by collecting rainwater and capturing fish and birds. At virtually any landfall, repairs could be made to vessels with resources to hand, and food and water were plentiful. But more than this, explorations were made to where human societies were already surviving in those environments and had even refined resources for trade or appropriation. The initial profits for the exploratory voyages to the New World came from already mined and refined gold and silver. Similarly voyages to the East relied on cultures that had already harvested and made ready for consumption the spices of the region. Even in adventurous explorations to extreme environments like the polar regions humans had free air and water, and even game to hunt as a backup to supplies they carried with them. The adventures of Colonists off the Earth have no analogy with human expansion over the Earth, and, as I discussed in an earlier paper (Kennedy 2013), will have none of the biosphere productivity implicit in the Earth economy as a free foundation to the costs of colonisation. Our colonists will be operating under the most severe constraints of sustainability and without the exponential growth of the natural world to supply basic needs. This backing to human growth from the integrated exponential growth inherent in the biosphere is the most disregarded calculation in efforts to imagine a sustainable settlement in space where there is no such foundation.

As Jared Diamond showed (Diamond 1997) the desire to move and explore is dependent upon a certain level of sustainability provided by the biosphere and where this sustainability is lacking societies do not expand. Moreover, any sustainability needs to embody the exponential growth factor of life if a society is to develop. The greater than replacement numbers in each generation is an essential requirement for a steady state minimum because of unexpected losses that arise from accidents and rises and falls of predators and resources. This last point is perhaps the most critical of all. It is precisely here that the distinction between the functions of growth is most in evidence. Exponential growth is damaging to the

Earth because it is wasted whereas exponential growth is essential for a space colony because simple linear growth will not allow it to survive in realms where there are no costless integrated autonomous systems to support every component.

14.4.3 Creating Colonies

Any individual component of life is a derivative function of the whole, and human life is no exception. The lack of natural environmental growth rates will have consequences not only on the effectiveness of solutions to the problems of life in space, be they technological or biological, but on the structure of any space society because linear growth leaves no room for error, or failure. In this respect, then there is no liberty in space in the sense of freedom of choice, because there will be no planned excess available to absorb options. Given the capitalist model, every activity in a space colony will be at the margin, and similarly dissent will have little or no excess with which it can shift to other means.

Thus, when it comes to space where there is no on-going support from the biosphere growth function nor any life on which to start building, successful colonisation will have to begin with a foundation that mimics such exponential support where possible. The particular game theory strategies of this reproductive principle of life that concern us are,

1. local maximum use of resources (i.e. no waste).
2. local maximum of resource availability in each trophic level while sufficient excess is available to all levels. (Unallocated resources available for use.)
3. repetition of sites to protect the system since one stands for all at any location the system covers.

From these principles we can draw some conclusions about how a successful colonising project would proceed whether or not the processes are conducted mostly by automatic and robotic systems namely,

1. **Complete modularity** in design features for all colonies and craft such that any craft can dock with any other craft or colony. This also means common flight communications, common navigation systems and protocols. The notion that each nation should create their own space systems without relation to others is as absurd as it is likely. Further the idea of patents and copyright will have no place in such an environment. Research and development will be shared freely and innovations would be distributed as upgrades.
2. **Complete free flows of resources** and people between colonies which implies common command and control systems (like computer software) and common language. This is a basic requirement of safety such that any group can come to the aid of any other with minimal interference. There is already an example in the commonly agreed language for aviation, but space systems will require much more than a common language of communication and where say, all

labels are in one language, they will need a common interpretation of problems, solutions and their implementations involving design details like wiring colours and mechanical operations and means of attachments. All airlocks should function the same way, for example; all energy generating systems should have the same controls, and so on.

3. **All settlements must be geared** towards producing stores of equipment and modular systems surpluses, and including labour, to be freely available to any other to fulfil the excess requirement for growth mentioned above. Any settlement should have personnel in excess of that strictly needed for planned work, and redundancy in automation.
4. **Facilities for care and hospitality** such that all groups can integrate freely and provide each other's replacements when in need, and most of all during unpredictable events. Let us accept too, that due to automation and artificial intelligence, a good part of colony life will be simply the voyaging and connecting with each other, and a 'touring of the boundaries'. The idea that colonies and the colonists in them will be operating like factory workers in the industrial revolution is archaic. Libertarian and dissenting behaviour will not be predicted upon that past.
5. **Free supplies** of oxygen, water, fuel and energy to be freely available to all (a component of every colony will be to contribute to the general stores of these resources above and beyond what they need for themselves) and sited at various points around the solar system with all the necessary systems to keep them viable.
6. **Complete openness** of data and decision-making processes. Without secrecy, society balances itself, and leadership is prevented from becoming *authority*.

This last point needs expansion. An initial criticism of this list of principles will undoubtedly be what has come to be called 'the tragedy of the commons'. Common resources get overused because selfish users compete to profit the most out of them (Hardin 1968). This portrayal of the users of a common-pool resource has been widely accepted as an indication that government structures are required to impose decisions since resource users are the least capable of managing the resources themselves. However, widespread collection of data shows this concept to be far too simplistic, and communities around the world do manage their commons effectively without authoritarian command and control where allowed to do so, as I can personally testify. Years ago I came to live in a (relatively) remote Basque farming community in the French Pyrenees. Much of the land in use is common land. Woods and summer grazing are all resources owned by no one (except the State) and are governed by the community. All the citizens have rights to the common resources (like wood and grazing) but use and distribution is governed by custom and practice as detailed in the anthropologist Sandra Ott's doctoral thesis about this very community (published in book form, Ott 1981). What is so interesting about the sharing customs is that where an individual behaves selfishly for some reason, or goes back on his word, there may be no sanctions at all. The individual selfishness is just absorbed and a spread of similar selfish behaviour does not result. An example described by Ott (*op.cit.*, pp. 146–147) of where two individuals of a team sharing work on the high pasture for their flocks of sheep decided to quit early. The others did not feel it right to continue to make cheese for themselves, the milk from the ewes was spilt on the ground, even though they were forced to look after the combined flock. There was no sanction against the two individuals. The brother of a neighbour to me, who lives on the coast and who returns to his natal house almost every weekend to hunt regardless of the season, has not been denounced by anyone, and the other hunters of the village who hunt in teams still obey the regulations. In game theory terms, in this community at least, defectors are trounced by generosity and defection does not spread.

A mountain village may be small scale, but colonies will be such small communities for a very long period of time. All the same, there is no reason why the open agreements between users applied at the point of use cannot be scaled up to larger communities. The millions of Twitter and Facebook users manage to apply narrowly focussed pressure on decision-makers without any centralised power structure at all. Theories of selfish behaviour derived from the Prisoner's Dilemma are entirely artificial since the heart of the problem is the matter of authoritarianism. Full information about the situation is withheld from the players who try to maximise their outcomes based on history or guesswork. The Prisoner's Dilemma is only a dilemma because there is an authority who has arranged the payoffs and keeps each player guessing about what the other has done. In the real world, cooperation does not need to arise in guesses about the other player's intentions. In fact, fuller game theory analysis shows that cooperation trumps every other strategy as long as there is full information shared between all the players. In this way the commons can easily be administered by self-interested groups who share all the information and where the 'obvious' solution becomes evident. Thus, the tragedy of the commons need not be an excuse for first creating authority to rule over a settlement, and there are alternatives which humans naturally create in the absence of a single sanction-based authority.

Regarding these principles listed above, it is important to realise that this implies a fundamental common society since no society can have special rules, beliefs or practices that would interfere with the flow between settlements. This would in principle rule out competitive expansion and extreme cultural divergences, often the source of dissent and restraint of liberty. The organisation of space settlements will necessarily be a counterculture of political uniformity not imposed from above but agreed upon among the individualistic elements that make them up.

A particular class of alternatives has been called 'polycentric governance' (Ostrom 2010) which consider that the socio-economic setting and more importantly the motivational climate can manage resources better than single authoritarian structures. Lon Fuller developed this idea of polycentricity to show that some disputes, resource allocation being a prime example, are inherently unsuitable for definitive adjudication due to the complex issues and interdependent interests involved, although he did believe in creating a 'rule of law' to which citizens should adhere (Fuller 1964). True polycentric decision-making, however, is more discretionary. Where webs of interdependent relationships are so complex that no external or anticipatory rule can be devised to extract a decision or solution, solving problems requires discretionary rules and local agreements that are realised to be non-binding forever but have time limitations on them. Polycentric decisions are made as inclusive as possible and that the grades of benefit to all parties are acceptable for the conditions as they stand but do not necessarily create precedents. They are also flexible in allocating benefits, such as cyclical bonuses, which are much more accepted by individuals than fixed allocations. Provisional allocations are less readily thought of as confrontations between classes or between privileged and non-privileged groups and can benefit political and economic decisions precisely because they are provisional. In space colonies where there will be a mix of humans, thinking machines and physically adjusted transhumans, there cannot be a set of rules established in advance that accommodates all views all the time. Polycentric decision making processes realise that fairness is mutable, and this removes an excuse for much dissent.

In fact humans are very good at working at complex tasks under these kinds of conditions of informed but non-preemptive decisions, and, when allowed, seem to naturally tend in this direction. The growing phenomenon of the civil society sector, called variously nonprofit, voluntary, the 'civil society', the 'third society', the 'social economy,' 'NGO,' or the 'charitable' sector, includes within it hospitals, universities, social clubs, professional organizations, day care centres, grassroots development organizations, health clinics, environmental groups, family counselling agencies, self-help groups, religious congregations, sports clubs, job training centres, human rights organizations, community associations, soup kitchens, homeless shelters, and many more. These are groups set up by private individuals to perform often vital services especially in health, education, and care as well as philanthropy. Each individual entity is generally small scale and are unique structures privately created by individuals coming together entirely on their own initiative but,

> ...Because of their unique combination of private structure and public purpose, their generally smaller scale, their connections to citizens, their flexibility, and their capacity to tap private initiative in support of public purposes, these organizations are being looked to increasingly to perform a number of critical functions: to help deliver vital human services, such as health, education, counselling, and aid to the poor, often in partnership with the state and the market; to empower the disadvantaged and bring unaddressed problems to public attention; to give expression to artistic, religious, cultural, ethnic, social, and recreational impulses; to build community and foster those bonds of trust and reciprocity that are necessary for political stability and economic prosperity; and generally to mobilize individual initiative in pursuit of the common good... (Salamon 2010)

These associations are not designed to make profits for investors save only to support whoever does the work and to return to the work as much of their income as possible. Nonprofit employment in the eight countries for which time-series data were available grew by an average of 24 %, or more than 4 % a year, between 1990 and 1995. By comparison, overall employment in these same countries grew during this same period by a considerably slower 8 %, or less than 2 % a year (Salamon 1993).

These are not new social inventions, but the conditions are right in the modern era for them to have a strong presence. They were noted in the US in 19th century by the observer Alexis de Tocqueville, who considered "…voluntary associations a uniquely democratic response to solving social problems…" It is tempting to think of their growth in the modern day is, too, a response to the loosening of the obligations of citizens to the State invoked by zero or negative interest rates we discussed earlier. While we are considering these social forms as more belonging to space settlements where the state influence is naturally less, it is the case that non-state actors are gaining control of features of everyday life from governments on Earth, and this includes the multi-faceted enterprises of the space economy. Since competition between nation states will only reduce the rate at which colonisation can proceed rather than assist it, deeper levels of cooperation and coordination, using the polycentric model of decision-making, between space development bodies on Earth will be needed. We can expect that NASA and other State organisations will become less and less important in the years to come and will be regarded only as components of a complex system of industrial and social effort in space. They will have authority in science perhaps but not necessarily in other areas of activity.

If we take, therefore, the existing adaptable polycentric collective model as the foundation to our social and economic structures in space, it only remains to find what economic instruments will support and encourage them, deliver preferable returns to the normal capitalist deployment of capital, and support both more rapid expansion into space while keeping to sustainable industrial and technological practices on Earth.

14.4.4 SpaceCoin

The economic instrument we are seeking will have to do four things, support polycentric decision-making, provide investment instruments, act as collateral and provide for wages and interpersonal currency among space settlements, while at the same time be as independent as possible of the fluctuations of fiat currencies.

So will colonies have money? How will individuals be paid and how will they store their income? Analogies with Earth-based trading voyages of the past may not provide a solution to the difficulty. Traditionally, merchant ship crews shared in the profits of the voyage in terms of a percentage of the value of the landed cargo. Navy crews received pay but also shared in the value of captured ships and other prize monies. It is unlikely that astronauts will have any opportunities to share in profits from space industries in similar ways in this century. Further, it is unlikely that human passengers in a space mission could contribute anything but a tiny fraction of the cost of such a mission so distributing shares in the equity to them makes little sense. Similarly the concept of 'sweat equity' where the labour that participants do during the lifetime of the project becomes their investment in it makes no sense when the costs of these voyages vastly outweigh what individuals have to contribute (unless, of course, all colonisers are billionaires) and would thus require that labour be re-evaluated enormously highly. 'Sweat equity' is perilously close to bonded labour and I do not imagine that it will have any serious role in space development. People have talked about setting up space banks, or using cryptocurrencies for space voyager payments. Some have proposed setting up a bank that offers interest free loans to stimulate space endeavours but this is no different to today's climate of zero or near-zero interest rates. It also misses the point about debt. The interest rate on its own cannot encourage particularly longer term activity if the returns from the activity for which the loan was taken out are too low. Furthermore since interest rate reflects risks in the returns, zero interest rate implies no growth. Gold does not serve the purpose of a universal currency; the general user cannot test its purity or easily subdivide it as its value rises, and most of the time it is undervalued with respect to other currencies. The problem with using Earth-based currencies is that interest earned on loans or dividends payable to share capital of a space venture simply cannot be calculated, which means that wages are unlikely to be fair, a possible cause of dissent. PayPal announced in 2013 that they were setting up PayPal Galactic in order to be ready to process payments that future space workers will make among themselves. This initiative seems to have since disappeared from the PayPal web site, with good reason. Such a proposal seriously misunderstands the space economy of this century. There will be no shops in space. Food and personal items will not be sold there. It will never be cost effective to ship and stock in space objects of anything other than essential food, hygiene items and perhaps clothing items. There might be room on re-supply missions for objects ordered from Earth, but the idea that a 'trading post' might be one of the components of a settlement on Mars is not practicable this century and 3-D printing would provide for many needs, in any event. So if astronauts do not buy and sell to and from each other, will they need money at all? Clearly they can buy and sell objects on Earth through electronic trading. So astronauts and colonists can be paid into Earth bank accounts and they can handle transactions relating to their life electronically from those accounts just like astronauts in the ISS do today. Colonists who have taken a one-way trip will presumably be paid either in a lump sum or a regular wages. They, too, are likely to prefer to put their money to use on Earth for the foreseeable future. But what can a colonist's life actually be worth? Certainly there will be declining value attached to each new colonist. Later arrivals will have different calculations applied to their 'pay'. These differentials can easily be causes of dissent. How can colonist wages, or indeed any space activity, be related fairly to the Earth economy?

There is such an economic tool already in existence namely, crypto currencies. The original purpose of a currency was to reflect the productivity of the community and to a certain extent its seriousness in paying debts. Since the world came off the gold standard, the value of a national currency has been divorced from actual values of productivity or precious metal holdings and has become related instead to trade balances which can be manipulated almost with impunity through interest rates. This has happened in part because, in exponential growth situations, productivity is a 'past measure' and cannot provide enough riskless future potential for investment or for the rising costs of government expenditure on wars and social services. As a result fiat currencies do not reflect true values in the economy. They are a construct and are vulnerable especially to a falling velocity of money.

There is a modern alternative to this namely, electronic currencies derived from the invention of blockchain verification and they all share similar features. The blockchain is in principle an unbreakable encrypted record of every single transaction (hence crypto-currency). The values in these electronic creations reside solely in the numbers of transactions made rather than the absolute values; its value deflates if people hoard it. This is reflected in the fact that coins are created every time the transaction list has to be verified after a transaction and which therefore contribute to its availability. Furthermore there is an actual limit to the whole numbers of coins that can be created so the subdividing of a cryptocurrency (essentially inflation) is in complete control of the users and is in direct relation to the need for it in a transaction. A transaction in a cryptocurrency does not pass through an intermediary like a bank, so there are no levies on top of the transaction inflating its value for the user, and there is no central control to manipulate its value. It makes no sense to have a debt in a cryptocurrency since a promise to pay cannot be recorded in the blockchain system, so debt cannot accumulate and alter its value. Yet there can be contracts to pay a sum in cryptocurrency bought with cryptocurrency. The difference between this contract to pay and a traditional fiat currency money note (I promise to pay the bearer…) is that the liability of the contract is not recorded in the blockchain only the fact that seller has received payment, and the contract has a term to expiry. Until the seller of the contract has acquired the coins of currency that he is expecting (say from the sales of goods) and paid off his contract does the multiple of the contract appear in the blockchain and thus in the value of the currency.[5] The contract is a legal instrument (it can be bought and sold) but its promises cannot be verified or accounted with reference to the block chain. Thus, in principle, a cryptocurrency value reflects something of the total concrete productivity of the arena in which the currency is used. It is this advantage in particular that is of interest to space development.

If the entire arena of space activity, including all the various groups and associations involved, from NASA and other national programs to the lowliest amateur group were to use a crypto-currency—let us call it SpaceCoin—then, as space activities rise over time, the value of the currency would also rise over time in direct relation to the total of transactions required. In this way SpaceCoin would reflect the value of both the industry and social activities, of scientific groups both professional and amateur, research and development, congresses and periodicals, tourist trips and so on. Any expansion in the space activity arena would increase the value of the currency automatically. Any resources discovered say, metal on an asteroid or alien microbes with genetic potential, sold in SpaceCoin would automatically raise the value of the currency for every holder. Each individual in the SpaceCoin economy would benefit without needing deposit accounts and interest rates, just as

[5]Bitcoin is therefore not the answer to debt problems of countries like Greece.

in the Feudal era, where a successful harvest profited everyone. Where in the capitalist system, borrowings can inflate shares leading to bubbles and collapses, with a cryptocurrency no bubbles of this sort are possible. SpaceCoin would be stable, capable of ever rising value but always be in proportion to the 'GDP' of the space economy arena. Eventually, any organisation wishing to enter the space development arena would prefer to use SpaceCoin for its stability rather than risk the variable values associated with fiat currencies. As more organisations buy into the arena the value of everyone's holding rises. But further, the most attractive aspect to SpaceCoin is the fact that since the background value to the currency is the numbers of transactions rather than any asset, asset losses in space would not move the value of the currency by large amounts since their value is already written into the currency (by way of the transactions) and cannot be removed.

As far as the Earth's economy is concerned this use of a cryptocurrency is, in effect, a return to feudal-like resource allocation where labour and capital are not distinguished in the wealth creation, and where waste is a loss for everyone. For the same reason that debt cannot be turned into an asset (outside of specific contractual arrangements), waste cannot be passed on to others. This makes the SpaceCoin arena a very attractive proposition for environmental groups and for sustainable industries. Members of the arena would find warfare a losing proposition and government spending could not exceed the resources available for it. As the benefits become clear to Earth-based groups one might expect a flight into SpaceCoin and away from fiat currency economies, braking destructive capitalist tendencies on the biosphere without any political domination.

So who creates SpaceCoin? Bitcoin, already in existence provides the model for SpaceCoin. BitCoin, an advance on earlier electronic currencies, was created in 2009 by a mysterious person or group of people who have remained crypto themselves, but all the software necessary to create the blockchain verification (mining) and the digital 'wallets' used to hold bitcoins is open source. There is no central agency controlling this process, only a group of dedicated individuals who run a web site to make available the software and who have created help forums to discuss its use. Bitcoin has a limit of 21 million units because the bitcoins generated by proving the blockchain are regularly halved such that there is a limit of 21 million reached by around the year 2140. The software allows for subdivisions of a bitcoin to the 8th decimal place ultimately producing 2100 trillion bitcoin fractions which compares roughly with a 2.5 % growth of world GDP to 2140 from $107 trillion today. There are theoretical risks to creating a cryptocurrency. For example if one miner or group of miners acquires more than 50 % of the mining capacity (this has already happened with BitCoin), they could in theory alter the blockchain in subtle ways to eliminate some transactions or repeat them. Some commentators say that this ability would actually depress the value of the currency and so automatically provide checks and balance to nullify such cheating. But even if not, the costs of doing these small 'cheats' continue to rise as the computing power needed to perform the blockchain verification rises. Like BitCoin, SpaceCoin would require an agency to write the software and run a distribution and troubleshooting. But where SpaceCoin would differ from BitCoin would be in its relation to the 'GDP' of the space economy and may require a larger projected quantity to take the space development economy through two centuries.

The UN seems to be the obvious choice to create SpaceCoin, although in fact it could be produced by any group of interested parties. SpaceCoin would give us,

1. A space economy that avoids the pitfalls of traditional and orthodox Earth-based economic management by government regulation of debt. One whose returns compete with traditional interest bearing instruments but whose value is generated apart from any fiat currency. An economic system built on value rather than debt.
2. A space economy that participates in the struggle to keep Earth's biosphere whole and as undamaged as possible. For example resource observation should be used to conserve biodiversity and not for developing successful monocultures.
3. A space economy that fosters mutual cooperation and horizontal decision-making rather than as simply a further outpost for government controlled asset allocation and capitalist waste economies.
4. Further initiatives to the study and management of space as a resource to preserve and enhance the Earth's Biosphere, as distinct to thinking of space just a common good to exploit apart from it. To clarify the treaties that recognise the commons of space and push for a new one that confirms the protection of Earth first.

14.5 Conclusions

The world economy is undergoing systemic changes in the relationship of capital with labour. Demographic changes and the political decoupling of capital from labour have led to an appetite for higher investment returns from the use of capital, and where labour no longer has a reasonable share in wealth derived from its mobilisation. Traditional enterprises in manufacturing sectors of the economy are more socially beneficial to labour, but new enterprises have emerged, aided by governments who have supported the debt-laden economies with rising asset growth over rising wages, which profit in particular from the details of individual consumer behaviour, including life and spending preferences, to manage their output more efficiently. A two-tier consumer world is being created which will actively hinder future innovation and do little to reduce the consumption that is damaging the world.

The implications of this are significant to the future of the Space Economy since its long timescales and lack of returns are very much less attractive to capitalists looking for the average higher returns now becoming the norm. The paradox is that capital will only turn to the space economy when either Earth's consumer economy becomes too damaged to provide the high returns for them, or when Earth economy has grown so large that any solution on Earth to substitute essential resources with other strategies do not work and it needs the resources the space economy can provide. It is unlikely that this point will ever be reached given the damage of such growth to the Earth's biosphere, and nor will a sustainable space economy be in place before this critical time since it will have to rely upon the purchasing power of Earth long into the future. A healthy Earth is necessary for any form of sustainable space economy to develop. Otherwise the current funding model of space activities where governments seed private enterprises to develop solutions to their requirements will leave a space economy pursing scientific studies and experimenting with almost whimsical colonising projects between the Earth-Moon system and Mars, without any coherent purpose except only an expectation that the Earth economy will catch up eventually and give these projects true purpose and an economic effectiveness. The investment multiple that supplies benefits here on Earth for every dollar of government funding will not continue to benefit the Earth economy unless the spin-offs are directed deliberately towards making the Earth sustainable.

This paper proposes a solution that will release the space economy from the restrictions that its dependency on old fashioned capitalist practices produces, and allow it to grow effectively without the biases of a central authority. I have shown that the consequences of high levels of debt around the world, a decline in labour as a consumer force, a decline in the sharing of capital, have led to a loss of authority of the State in controlling the economy, and a rise of the non-state actor in its place. I propose that in particular we give strength to the non-state actors by binding together all aspects of the space economy, industrial, social and spiritual, voluntary and professional, with a space economy cryptocurrency, SpaceCoin, in order to separate the space economy from the perils of this late stage capitalism, to quicken the pace of space development and to provide a non-political biased impetus to saving the Earth's biosphere. Such a program would mesh well with the UN's new initiative for sustainable development goals as defined in the United Nations Sustainable Development Summit of 2015 (UN 2015).

I conclude that these proposal will provide the appropriate growth foundations to future space colonisation, and, with polycentric decision-making and a space-wide currency in place, there will be few reasons to worry about liberty, dissent and revolution in future colonies that will naturally grow in value as the space economy grows in value.

We shall in effect be making a quasi-feudal 'space fief' where labour and capital join together to save the Earth and make a home for humans in space.

14.6 Final Word

We have a strange sense of risk in the western world. It has two components. One notion, to which we give investment value and which has come to create and dominate our modern monetary-based society, is the relationship of monetary gain to risk; the greater the risk greater the return. But this is far from a world-wide appreciation of risk. The other notion, which we celebrate more where we can, is the human individualism in risk, where someone risks all for personal fulfilment. Wade Davis mentions Polynesian historians who estimate that half of all Polynesian voyages risked in the Pacific were lost (Davis 2009). That is perhaps the finest expression of risk—the toss of a coin. The Polynesian adventurers were not seeking their own wealth or even to set up long lasting trade lines between islands. They went to express themselves; they went for status; they also went to confirm the knowledge that other human settlements, although dispersed, lived within reach, and even though well beyond the horizon, those settlements could be bound into one sphere by a voyage of risk. The whole community was involved in the voyage, provided the resources, and the whole community gloried in its success. This expression of risk seems the most adapted to the Space adventure, where settlements will communicate with each for little reason but to confirm their presence in the life stream, their extended family bond, their humanity. The future of humans in space should not be interpreted as an escape path from a dying Earth or as simply a natural extension of human economic and political structures, but as an expression of what humans value most about all human life. Space is ultimately not about Earth's economics but about human consciousness and about the experiences the underlying cosmic realities of life induce within the whole human community. It only has meaning with respect to our origins.

References

Amundson, R., Berhe, A. A., Hopmans, J. W., Olson, C., Sztein, A., & Sparks, D. L. (2015). Soil and human security in the 21st century. *Science*. doi:10.1126/science.1261071.

Angell, I. (2000). *The new barbarian manifesto: How to survive the information age*. London: Kogan Page Business Books.

Beckert, S. (2014). *Empire of cotton: A global history*. New York: Knopf.

Bolton, J. (2012). *Money in the medieval English economy: 973–1489*. Manchester: Manchester University Press.

Brand, S. (Ed.). (1977). *Space colonies* (p. 86). London: Penguin Books.

Chapman Research Group. (1989, June). An Exploration of Benefits From NASA. *Spinoff*.

Chase Econometric Associates. (1977). *The Economic Impact of NASA R&D Spending: PAD-78-18*. US Government Accountability Office.

CIA World Fact Book. (2015). *Spending on space by country*. https://www.cia.gov/library/publications/resources/the-world-factbook/. Accessed May 21, 2015.

Cockell, C. (2013). *Extraterrestrial liberty: An enquiry into the nature and causes of tyrannical government beyond the earth*. Edinburgh: Shoving Leopard.

Confessore, N., Cohen, S., & Yourish, K. (2015, October 10). Buying power. *New York Times*.

Darrat, A., & Al-Yousif, Y. (1999). On the long run relationship between population and economic growth: Some time series evidence for developing countries. *Eastern Economic Journal*., *25*, 3.

Davis, D., et al. (2004). Changes in USDA food composition data for 43 garden crops, 1950 to 1999. *Journal of the American College of Nutrition*, *23*, 669–682.

Davis, W. (2009). *The wayfinders, why ancient wisdom matters in the modern world*. Toronto: House of Anansi.

Diamond, J. (1997). *Guns germs and steel: The fates of human societies*. New York: W.W. Norton.

Dyer, C. (2002). *Making a living in the middle ages: The people of Britain 850-1520*. New Haven: Yale University Press.
Economist. (2013, December 16). Inequality: A defining issue for poor people. *The Economist*.
Economist. (2014, November 1). Business in the blood. *The Economist*.
Economist. (2015, April 16). Family companies, to have and to hold. *The Economist*.
European Commission. (2015). Bringing down barriers in the Digital single market: No roaming charges as of June 2017. *European Commission Press Release database*. http://europa.eu/rapid/press-release_IP-15-5927_en.htm. Accessed October, 27 2015.
European Space Agency. (2015). Web site. http://www.esa.int/Our_Activities/Space_Engineering_Technology/A_solid_investment. Accessed May 2015.
Ferguson, C. K., et al. (2015). Space 2100: A shared visioning exercise for the future space economy. *Journal of the British Interplanetary Society,* 68, 10–16.
Ford, M. (2015). *Rise of the robots: Technology and the threat of a jobless future*. New York: Basic Books.
Fuller, L. (1964). *The morality of law*. New Haven: Yale University Press.
Graedel, T. E., et al. (2011). *Estimating long-run geological stocks of metals, working paper*. UNEP International Panel on Sustainable Resource Management Working Group on Geological Stocks of Metals, 6 April.
Greenpeace Research Laboratories. (2008). *Dead zones*. UK: University of Exeter.
Gurtuna, O. (2013). *Fundamentals of space business and economics*. New York: Springer Press.
Hardin, G. (1968). The tragedy of the commons. *Science, 162*, 1243–1248.
Hawking, S. (2010). Stephen Hawking: Mankind must colonise space or die out. http://www.theguardian.com/science/2010/aug/09/stephen-hawking-human-race-colonise-space. Accessed January 06, 2015.
Hurun Research Institute. (2015). Hurun global rich list 2015:February. http://www.hurun.net/en/articleshow.aspx?nid=9607. Accessed May 2015.
Impey, C. (2015). *Beyond: Our future in space*. New York: W. W. Norton & Company.
International Fertilizer Development Centre. (2010). *Report: World phosphate reserves and resources*. Alabama: IFDC.
International Labour Organisation. (2014). *Profits and poverty: The economics of forced labour*. Geneva: International Labour Organisation Publications.
IPCC Working Group II Report. (2014). *Climate change 2014: Impacts, adaptation, and vulnerability. Fifth assessment*. World Meteorological Organisation: Geneva.
Kennedy, A. (2013). Interstellar travel: The broader consequences of the minimum time to interstellar destinations from now and its significance to the space economy. *Journal of the British Interplanetary Society, 66*, 96–109.
Kharas, H., & Gertz, G. (2010). The new global middle class: A cross-over from west to east. In Cheng, L. (Ed.). *China's emerging middle class: Beyond economic transformation* (Vol. 2, pp. 18–22). Washington: Brookings Institution Press.
Lyttle, D. (1991, February 6). Is space our destiny? *Astronomy*.
Marx, K. (1867). *The origins and development of capitalism*.
Messier, D. (2014, January 14). Shotwell: Reusable falcon 9 would cost $5 to $7 million per launch. *Parabolic Arc*.
Miller, C., et al. (2015). Economic assessment and systems analysis of an evolvable lunar architecture that leverages commercial space capabilities and public-private-partnerships. Nextgen Space. http://www.nss.org/docs/EvolvableLunarArchitecture.pdf. Accessed May 20, 2015.
Moore, A. (2015, February 15). Negative interest rates to shake up financial system. *Financial Times*.
Mumford, L. (1938). *The culture of cities* (pp. 42–45). London: Secker and Warburg.
Musk, M. (2004). Aeon magazine, 30 september. http://aeon.co/magazine/technology/the-elon-musk-interview-on-mars/. Accessed January 06, 2015.

Nagelkerken, I., & Connell, S. D. (2015). Global alteration of ocean ecosystem functioning due to increasing human CO_2 emissions. In *Proceedings of the National Academy of Sciences*, October 12.
NASA. (1977). *Space settlements—A design study*. NASA SP-413. USA: NASA Publications.
NASA. (2015a). Web resource. *Spinoffs*. https://spinoff.nasa.gov. Accessed May 2015.
NASA. (2015b). *Journey to mars: Pioneering next steps in space exploration. NP-2015-08-2018-HQ*. USA: NASA Publications.
Odlyzko, A. (2003). *Privacy, economics, and price discrimination on the internet [Extended Abstract], revised July 27*. Digital Technology Center, University of Minnesota. http://www.dtc.umn.edu/~odlyzko/doc/privacy.economics.pdf. Accessed October 1, 2015.
OECD. (2011). The space economy at a Glance 2011. *Organisation for Economic Co-operation and Development*. http://www.oecd.org/sti/futures/space/48301203.pdf. Accessed May 10, 2015.
Ostrom, E. (2010). Beyond markets and states: Polycentric governance of complex economic systems. *American Economic Review 2*, 1–12.
Ott, S. (1981). *The circle of the mountains*. Oxford: Clarendon Press.
Salamon, L. M. (1993). *The global associational revolution: The rise of the third sector on the world scene* (p. 29). Occasional paper: Johns Hopkins University Institute for Policy Studies.
Salamon, L. M. (2010). Putting the civil society sector on the economic map of the world. *Annals of Public and Cooperative Economics, 81*(2), 167–210.
Sample, I., & Branigan, T. (2011, April 26). China unveils rival to international space station. *The Guardian*.
Schmoch, U., et al. (1991). Analysis of technical spin-off effect of space-related R&D by means of patent indicators. *Acta Astronautica, 24*, 353–364.
Schor, J. (1999). *The new politics of consumption: Why Americans want so much more than they need*. Summer: Boston Review.
Singh, K. (2009). *Rural development: Principles, policies and management* (3rd ed.). New York: Sage Publications.
Singman, J., & McLean, W. (1999). *Daily life in medieval England*. Westport: Greenwood Press.
Standing, G. (2014). *The precariat*. London: Bloomsbury Academic.
The Oxfam issue briefing. (2015). Wealth: having it all and Wanting more. Oxfam GB. https://www.oxfam.org/en/research/wealth-having-it-all-and-wanting-more. Accessed September 1, 2015.
The White House. (2012). *Fact sheet: The national space policy*, 28 June, 2012. https://www.whitehouse.gov/the-press-office/fact-sheet-national-space-policy&langpair=1%2C2. Accessed January 10, 2015.
UK Office for National Statistics. (2015). Summary of labour market statistics, October.
UN Department of Economic Affairs, Population Division. (2015). *World population prospects, the 2015 revision*. United Nations. http://esa.un.org/unpd/wpp/. Accessed August 8, 2015.
United Nations. (2014, December 5). Spotlighting humanity's 'silent ally,' UN launches 2015 International Year of Soils. *UN News Centre*. http://www.un.org/apps/news/story.asp?NewsID=49520#.VjeLjYSfxTo. Accessed May 20, 2015.
United Nations. (2015). Transforming our world: The 2030 Agenda for sustainable development. A/RES/70/1. United Nations.
US Bureau of Labour Statistics. (2014). *Current population survey, July*. http://www.bls.gov/cps/_23data. Accessed May 15, 2015.
US Bureau of Statistics. (2014). *Databases, tables and calculators by subject*. http://www.bls.gov/data/. Accessed May 15, 2015.
US Commission on Space. (1986). *Pioneering the space frontier*. New York: Bantam Books.
Vaccari, D. A. (2009). Phosphorus famine: A looming crisis. *Scientific American*. 54–59.
Weeden, B. (2012, June 4). The economics of space sustainability, *The Space Review*.
Whitesides, G. (2010, November 10). George.Whitesides, Chief Executive, Virgin Galactic. *SpaceNews*. http://spacenews.com/george-whitesides-chief-executive-virgin-galactic/#sthash.LhPGk74g.dpuf. Accessed May 2015.

Willamson, S., & Cain, L. (2011). *Measuring slavery in 2011 dollars*. http://measuringworth.com/slavery.php. Accessed June 25, 2015.
Wood, E. M. (2002). *The origin of capitalism: A longer view* (p. 8). New York:Verso.
Worldwatch. (2013). State of consumption today. http://www.worldwatch.org/node/810. Accessed August 8, 2015.
Yusingco, M. H. (2015, August 11). COMMENTARY: Political dynasties vs. political rights. *Mindanews,* http://www.mindanews.com/mindaviews/2015/08/11/commentary-political-dynasties-vs-political-rights/. Accessed September 1, 2015.

Index

C.S. Cockell (ed.), *Dissent, Revolution and Liberty Beyond Earth*,
Space and Society, DOI 10.1007/978-3-319-29349-3

MIX
Papier aus verantwortungsvollen Quellen
Paper from responsible sources
FSC® C105338

If you have any concerns about our products,
you can contact us on
ProductSafety@springernature.com

In case Publisher is established outside the EU,
the EU authorized representative is:
Springer Nature Customer Service Center GmbH
Europaplatz 3, 69115 Heidelberg, Germany

Printed by Libri Plureos GmbH
in Hamburg, Germany